The Deadly
Element

The Deadly

THE STORY OF URANIUM

Element

By LENNARD BICKEL

STEIN AND DAY/*Publishers*/New York

First published in 1979
Copyright © 1979 by Lennard Bickel
All rights reserved
Designed by David Miller
Printed in the United States of America
Stein and Day/*Publishers*/Scarborough House
Briarcliff Manor, N.Y. 10510

Library of Congress Cataloging in Publication Data

Bickel, Lennard.
    The deadly element.
    Bibliography: p. 302
    Includes index.
    1. Uranium.   I. Title.
QD181.U7B52   669'.952'931   78-66243
ISBN 0-8128-2589-6

to FRANK DEVINE

There has been born the most
revolutionary force since
prehistoric man discovered fire.
ALBERT EINSTEIN (1945)

CONTENTS

ILLUSTRATIONS

FOREBODING

**STOCKHOLM—
1903.**

Is it right to probe so deeply into Nature's secrets? The question must here be raised whether it will benefit mankind, or whether the knowledge will be harmful. Radium could be very dangerous in criminal hands. Alfred Nobel's discoveries are characteristic; powerful explosives can help men perform admirable tasks. They are also a means to terrible destruction in the hands of the great criminals who lead peoples to war. . . .

PIERRE CURIE, *Nobel Prize Oration*

**CHARTWELL, KENT:
ENGLAND—1952.**

To avert a vast, indefinite butchery, to bring the war to an end, to give peace to the world, to lay healing hands on its tortured peoples by a manifestation of overwhelming power—at the cost of a few explosions—seemed after all our toils and perils a miracle of deliverance. . . .

WINSTON CHURCHILL, *Triumph and Tragedy*

MANIFESTATION

Life on earth is under threat—from a single element. That element is the trigger to nuclear arsenals of incredible strike power, and so has raised the prospect of self-destruction for mankind.

This statement is not to peddle fear—it simply declares the fact which mankind now faces. It is a situation grown from the discovery of this one element—uranium.

Locked within the complex and compacted cores of the atoms that form uranium are primordial energies which have been there since Creation, and which, when liberated, are alien to human senses; yet these are the natural forces which compose all matter, even the atoms of our bodies. Thus, by itself, uranium is not unnatural. It is human manipulation that makes this element the key to colossal explosions and the primer to by-products which now arouse universal dread. It is human folly that has bred the psychosis which makes its name synonymous with our planet laid waste, so that it looms over the future like the mushroom cloud of its own disintegration.

At the same time, ironically, uranium offers energy sources thousands of times greater than the dwindling fossil fuels over which energy-hungry nations could come to conflict—if there were no nuclear alternative.

That situation has brought mankind face to face with its greatest dilemma; it makes it critically important to understand the nature of this fundamental energy, which offers the alternatives of boundless power supply or nuclear devastation; and to ask, what is its history?

From knowledge made possible by the discovery of uranium, astrophysics is now close to explaining the mystery of Creation. It is a new Genesis grown from the concept glimpsed earlier this century by the Belgian astronomer-priest-philosopher, Georges Lemaitre: "The evolution of the world can be likened to a display of fireworks just ended—some few red wisps, smoke, ashes. Standing on a well-chilled cinder, we see the slow fading of the suns and try to recall the vanished brilliance of the origin of the world."

If this Big Bang theory is correct, then the energy liberated from the uranium atom is a leftover from that "vanished brilliance." In that unimaginable inferno primordial energy was locked into the fabrication of matter, into the first simple atoms, the hydrogen atoms which formed the gases, which formed the vast swirling clouds in which the first stars in our galaxy were born. Giant suns, unstable, many of them monster stars, exploded and imploded, again and again, and formed the synthesis of more complex atoms.

Astrophysicists now judge the age of a star by its metal composition; they can say our sun is 5 billion years old, perhaps half the age of the Galaxy; that the stellar synthesis of heavy elements has been under way for 5,000 million years—at least. Thus, out of old giant suns, long ago vanished, came the range of elements that form our world; and the most complex, power-packed atom was uranium.

This element was unknown for most of human history. Even when it was discovered, more than a century passed before its powers were sensed. By manipulation, men (fearing its use in the hands of a common enemy) finally induced it to display its violence across the arid desert of New Mexico. Puny against the exploding suns which created the element's structure, it was, yet, unlike anything seen on earth before . . . it was climacteric for man.

NEW MEXICO—
July 16, 1945.

The desert is quiet. Summer rain clouds that caused delay have fled; fine lines of lightning lick distant ranges, too far off to show a gleam to the dark, lonely landscape. Stars glimmer through lace clouds above the mesa near Oscura Peak and over the stretch Indians knew as the Valley of Fires. In the predawn

gloom the 100-foot-tall steel tower, with its grossly bulbous can-
nister suspended above the ground, is not visible; the firing cables, the instrument leads running from the control bunker—six
miles distant—are lost to view a few yards away.

The delay has increased suspense. Scientists, soldiers, technicians, observers, are already under stress. Suddenly, lights mark
the distant tower; the last five men at the firing point are racing
back to the control bunker. The loudspeakers crackle over the
mesa; the firing will take place at 5:30 A.M. In minutes they will
see the first contrived atomic explosion which—for an inexplica-
ble reason—has been code-named Trinity.

The countdown begins.

Groups of men—six miles, ten miles, twenty miles from the
firing point—lie on the ground with feet toward the tower. Some
just close their eyes, others wear dark-glass goggles, others peer
through specially treated glass. Then, the second is here; the
voice calls; with sharp excitement, *"Now."*

Six miles away, the bulky steel sphere hides the arranged
segments of silvery, glossy metal fashioned out of nature's heaviest element. These segments are like cuts from a melon; they are
surrounded by chemical charges timed to fire simultaneously,
cunningly placed to blast the melon shapes inward with great
force. The first reaction to the thrown switch is to form a whole
melon shape, of dense, imploded metal—a compact mass, a critical mass.

In much, much less than a billionth of a second, neutrons are
slicing through fierce fields of opposing and binding forces, into
packed cores of the element's atoms, there sparking the tiny
explosions that chain-react. At the heart of the metal melon,
across a volume of two cubic centimeters where there is no escape
for the hurtling neutrons, is the beginning of the inferno. Faster
than lightning the chain reaction flashes through crowded nuclei
in the dense metal, erupting in a hailstorm of flying neutrons,
fission fragments from shattered nuclear structures, and the unleashed, lethal gamma rays. All this released energy fulminates in
a fury of heat fiercer than at the sun's core, an explosive force a
billion times greater than the crush of earth's atmosphere: heat
and pressure no man-made container can resist.

In that first billionth of a second, cosmic energy bursts in free
release into earth's atmosphere: a fearful, ersatz star contrived by

human agency, an earth-bound sun which vaporizes all surrounding metal structures and materials and which, in an instant, fires the sands of the desert to glass—fused silicon that
afterward would be fixed in small samples into blocks of clear
plastic and labeled Trinitite.

Over the stricken desert flashes a light that pierces dark glass
and causes the men momentary blindness; a light that illumines
mountains a hundred miles away, a light that fills the dark sky.
Behind the quivering glare the monstrous fireball is forming,
which lifts its spreading mass in a fiery shroud of crimson, red,
orange, spuming scarlet, within a purple, violet halo. As it grows
a parasol stalk of swirling radioactive steam and dust, it lifts its
head into the first form of the mushroom-shape cloud.

It seems to dull slightly, but outward-streaming energy licks
wisps of moist cloud into traces of sudden white. And all in
silence—so far. Then comes a great surging wave of air, a blasting
shock front which bludgeons eardrums, which thunders and
echoes through the San Andres Mountains with the noise of a
thousand caravans reverberating in the flinty hills.

In Alamogordo, 60 miles to the south of Trinity, a false dawn
floods the heavens; in Carrizozo and White Oaks 40 miles eastward the sun seems to rise on the wrong side of the world.

From the fission of a few cubic centimeters at the innermost
heart of the imploded metal has come the mightiest explosion
achieved by men.

The explosion is so far in excess of the expected that, in the
bunkers, there is at first wild disbelief in what they have seen and
heard; soon, they are on their feet, cheering, grinning—the army
chiefs, the scientists, the technical assistants—all of them slapping
backs, shaking hands.

For the U.S. Army's second-in-command of the project, General Thomas F. Farrell, the sight and sound of the first atomic
explosion warns of Doomsday; that men are "blasphemous to
dare tamper with forces heretofore reserved for the Almighty."

A man, in charge of the firing of the explosion, is moved to
cry, "Now we are all sons of bitches!"

For others, other meanings. It will end the bloodiest war in
history; it will bring Japan to her knees and avoid the slaughter
of mass invasion. For others still, it will end the years of angering
doubts, the secret deals with uncomprehending political and
military leaders.

For the military establishment, it will give global supremacy to American arms and will validate the spending of two billion dollars on the building of a nuclear colossus to create the fissile explosives.

For the scientists, it will be recompense for frustrating bottlenecks, disbelief, the slights of obdurate administrators.

For the dreamers, there is a splendid vision—of a new kind of world for mankind in which there is access to a new source of unlimited power; science can now lift the crushing burdens of poverty, hunger, privation, and ignorance from the underprivileged millions and endow a whole fresh future for the war-torn world.

It will vindicate those scientists who fled dictatorship in Europe, and fought, argued, cajoled, sacrificed their time and prostituted their talents to create the world's most inhuman weapon to curb tyranny.

The nuclear fire on the mesa of New Mexico is hailed by many as the birth of the atomic age. But they are wrong. The cosmic eruption they have witnessed is an exposition of the human paradox: superb intellectual exploration, exquisite science being used for slaughter—because giant advances in knowledge coincide with conflict.

The birth of the atomic age occurred long before this day of the explosion; it was in the discovery of uranium, the heaviest substance in nature, number 92 in the periodic table of the elements; in the revelation of its properties and its astonishing construction, by men and women dedicated to probing the secrets of matter. Often ignored, underpaid, neglected, working in cellars, and damp basements, old sheds, they were driven by the urge to know the unknown. Their work has brought great benefits to man—and the great nuclear dilemma. This book is their story.

FIRST PHASE
Journey
Into The Unknown

BERLIN—
1780

Arm in arm they stand in the February wind. Both of medium height, they are wrapped in the usual dark, head-to-heels cloaks worn by Berliners to defeat the cutting northern blasts.

They stand and gaze at the shop, at Spandauer Strasse, 17. It has a signboard that reads

APOTHEKE ZUM GOLDENEN BÄREN

The paint is drab, and it looks run-down. Behind the windows the shelves are almost bare, and Christiane Sophie Lehmann knows the rooms in the upstairs living quarters will be dingy, colorless. At 32, she is content to part with her dowry of 9,000 silver thalers when they are wed on the morrow—on Friday, February 13. She will get an industrious, clever husband in Martin Heinrich Klaproth; and hard work and enterprise will bring their rewards.

He is now 37, with many years of hard work behind him. Tomorrow he will get a wife—and his own business. More than that; at the side of the shop in Spandauer Strasse is room for the laboratory where he can follow the interests he has built up since his youth.

Born in the medieval town of Wernigerode in the Harz Mountains, Martin Klaproth, one of three sons of a poor tailor, grew up in a narrow cottage in the shadow of the feudal stronghold of Count Christian Ernst von Stolberg.

Life was bitterly frugal, cold in winter from lack of fuel, and uneventful—until Martin was seven years old.

Fire swept the small town and left the little timbered houses smoking debris; the seven-year war with France surged back and forth over the area. Martin's father, Johann Klaproth, managed to place his three sons in a Latin college to study for the church, but, at 15, Martin was expelled for lack of funds. With only a year to graduation, he was sent to Quedlinburg and sold as an apprentice to the local apothecary.

Working for his keep and tuition, Martin Klaproth slaved during the hours of light, mashing roots with a wooden bludgeon, feeding open fires beneath steaming cauldrons, pumping bellows, running errands and delivering prescriptions at all hours in all weather, ill clad, ill fed—sleeping at the back of the shop on a straw palliasse. Most of his notes on these hard times were destroyed during World War II, in Berlin, but one surviving memo underscores the indifference, and neglect: "I am unable to boast of any tutoring from my master. I have to be satisfied by learning from others, by watching what they do, and by reading the outdated chemistry books, for which little spare time is allowed."

In 1763 he won a junior post with a pharmacy in Hanover and, after two years there—during which time he taught himself Greek to add to his Latin—he went to Berlin. He was so well liked by his employer that he was appointed, at 25, as executor to manage the business and take care of the widow and children when the owner died.

Klaproth burned with a zeal for reform, for better conditions for apprentices, for a sweeping away of quacks and charlatans from the ranks of chemistry. He showed that a secret panacea called "nerve drops"—for the recipe of which the Empress Catherine II paid a fortune of 3,000 rubles—was nothing more than iron chloride dissolved in ether.

He was a pioneer in true chemistry. He opened new paths from experimentation in his own laboratory so that he became a master of separating out single elements from the many compounds that came into his hands.

His collecting of various chemicals and minerals was assiduous and constant, and from these he extracted unknown elements—beryllium, the bright, brittle tellurium metal, chrome, cerium, and zirconium.

By 1789 his business and work has prospered, and he is examining samples sent by Baron Alexander von Humboldt, some from John Hawkins of Cornwall, England, and many from his own mineral collection.

He is studying what some men have called "pitchblende," others "black tin ore." The mineralogist A. G. Werner, of Freiberg, thinks it is "iron pitch ore." It is also known generally as a compound of zinc, tungsten, and iron.

One sample was sent to him from the George Wagsfort Mine at Johanngeorgenstadt, in the Erz Mountains, but he has a similar ore from Joachimsthal, in Bohemia. And he remembers seeing this deposit, with black nodules suggesting a mineral, resting on a bed of debris colored green-yellow and ochre.

He is careful to grind the ore into a fine powder, for that will increase its surface area and expose it to the processes and reactions he has planned. He is founding basic chemical procedures that will continue into the future. He uses sulphuric acid on his powder and tries to heat it with an apothecary's blow-pipe and a spirit lamp, but his breath does not fan the flame to a high enough temperature; he uses more powerful aids, turning to a mixture of nitric and hydrochloric acids and then heating the solution in his porcelain furnace in the most intense heat he can achieve.

On cooling—a slow, easy cooling without blowing air over the crucible—he finds the precipitate *is* tiny metal beads, beads of lead chloride. *And what is lead doing with the iron, tungsten, the zinc?* It is a question that will not find a clear answer for more than a century.

He works again, and again, on the mother liquid—the tests go on for weeks. Then, after using many of his assay techniques, he cools his solution slowly one evening, and, as he watches, entranced, crystals form under his eyes, a type of crystal he has not seen before, regular in design—four-sided, column-shaped, about half an inch long, of clear yellow.

In tests on the crystals, he obtains differing colors, and then he dries the liquid into soluble residue. He retraces his path, using the trick of oxidizing the material with acids, winning back his yellow salts. He produces acetates, phosphates, and sulphates, and from this assumes he has a material very much like metal. Yet he cannot reach the metal he suspects is in the powder. He mixies it with coal dust and borax, to provide additional heat and carbon, and cooks his material before adding linseed oil.

With patience, he finds his pathway to what he calls a "calcin substance"—it is a yellow oxide which seems to cake as it cools.

One night he turns back to the residue from the Joachimsthal mine; he subjects this to fractionation and finds it contains iron oxide, some lead, copper, and silicon. What, then, can this new substance be, this little heap of yellow powder lying in his dish? Surely it, too, must be metal? Again, with wax, with oil, he bakes a sample of 50 grains in the furnace. It produces little blobs of porous, spongy metallic matter. He uses another 50 grains, this time with acid, and wax and oil, and he increases the heat treatment.

Martin Klaproth finds a metal granule of about 28 grains. He tests it with a file—it shows a gray, metallic sheen where he cuts it. It *is* a metal—but different. Long before the world has any notion of atomic mass, he decides that this material has a high specific weight. He has come to the end of this trail.

His cloth coat impregnated with pungent fumes of acid and smoke, his eyes bloodshot, his head aching, his mind racing with the mystery still unsolved, he climbs the wood stairs to his waiting supper.

As he eats he tells Christiane, "I have only just found it, but I'm sure it's a new element. It's nothing to do with zinc, iron, or tungsten—or silver. Nothing like any substance I know. I think it's a strange kind of half-metal."

On the evening of September 24, 1789, Klaproth addressed the Royal Prussian Academy of Science, in Berlin:

> The number of known metals has been increased by one—from 17 to 18. This I have called a metalloid, a new element which I see as a strange kind of half-metal. It is not related to the zinc, iron, or tungsten found in so-called pitchblende. For some reason, however, I have found it associated with lead. Consequently, I suggest that past errors in naming should be eradicated—such as iron pitch ore, pitchblende, or black tin ore. I have chosen a name. A few years ago we thrilled to hear of the discovery of the final planet by Sir William Herschel. He called this new member of our solar system Uranus. I propose to borrow from the honor of that great discovery and call this new element—Uranium.

Seven decades before Mendeleev grouped the elements in the order of their nature and properties, to be called the periodic table, and more than a century before it was known that atomic mass (or weight) dictated their places in that famous scale, Klaproth sensed he had found the final element in nature.

But the element he revealed led to the advance in knowledge of the structure of matter, of the relationship between atoms and all living things; it made possible astounding technical advances, in communications, medicine, and astronomy—and warfare.

RUSSIA—1872.
A Train on the Steppe

In any crowd, at any time, in any country in the world, Dmitri Ivanovitch Mendeleev would have been noticed. In the third-class carriage of the train rocking across the steppe in the heat of July, 1872, he was a towering personality. With the verve and flair of Mongol blood, inherited from his Tartar mother, with deep-set, vivid, blue eyes flashing and massive head wagging and shaking for emphasis, he talked and ate black bread and pickles with itinerant farm workers and peasants. Regaling his companions with earthy stories of his boyhood in Siberia, he told of his life as the youngest of 14 children, of how his teacher father had gone blind and his enterprising mother had founded and managed a glassworks.

Mendeleev was thirty-five. He wore his hair in flowing locks, which he allowed to be trimmed only in spring each year; he shunned all adornments and decorations, and fervently espoused the cause of women. He was taking to St. Petersburg the first chart of the Mendeleev periodic table of the elements.

Educated by his blind father at a small Siberian school, Mendeleev had found a method of relating atoms to the elements. This monumental work was to throw new light on the variable character of matter by a brilliant arrangement of all the world's elements into groups. For the first time, his table gave ordered places to the elements, from the lightest—hydrogen, and helium—up the scale to the heaviest, and that element was uranium.

His periodic table would hang in almost every laboratory and every teaching college in the world. Yet after delivering his paper, Mendeleev went back home on the train, third-class, to resume his task as supervisor of weights and measures, to be totally ignored by the Russian scientific elite, to while away his years with oil painting and talking, unaware that his work and genius would play a role in an explosion of knowledge that came with the birth of the twentieth century.

WÜRZBURG, GERMANY—
November, 1895

In the physics building at the University of Würzburg, the professor of physics, Dr. Wilhelm Roentgen, is about to endow world healing with its greatest diagnostic tool. He is a small man. He likes to work alone and is secretive about what he does. He has a device known as a Crookes tube, built like a big light bulb lying on its side. It is fed with electric power from a high-voltage induction coil, and, in a sense, it is a forerunner to the television age; however, this night's work will have much greater impact.

Men have worked with such a tube across Europe, in England, and have discussed the detection of emanations (which have been called "ether" rays) that appear when the current is running through the gas in the big tube. Some have argued that these rays are like the glow from substances that gleam in the dark, that they are stored energy, ultraviolet light from the sun trapped between structures and released by some unknown mechanism. Roentgen doesn't like that. He has covered his tube with black cloth to shut out all light. He switches on the power supply. Quite unrelated to this experiment is a sheet of cardboard coated by chance with a special preparation of barium salts; and in that darkened room, as the power surges through the well-covered cathode tube, this chemical coating starts to glow. Roentgen moves it further away—two, then three meters—and still it glows. And there is no other light in the room.

Not daring to come to a conclusion, indeed, unable to attempt an explanation at this stage, he works into the small hours, varying, changing the cardboard, preparing another one with a preparation of barium platinocyanide. Always, the mysterious, unseen power streams across the room, from some action of the electricity on the anode and the glass. Invisible power—from nowhere! This, surely, has no relationship to stored ultraviolet light from the sun; this is totally different from anything known.

A week later, he calls his wife to his laboratory: "I want to show you something wonderful, and mysterious." In the darkened room Roentgen places his wife's left hand over a photographic plate still wrapped in protective lightproof black paper.

While the plate develops, he explains to his wife the experiments carried out each night behind the locked door of the laboratory: "Through sheets of tinfoil, through flesh, even through a thick book with 1,000 pages—straight through all of these—the unseen waves go. They are not deflected, not stopped. Even through a whole pack of playing cards—invisible power passing through solid objects!"

He warns her, "It is an historic discovery, and you must not breathe a word to a soul until my paper is written and published."

When the glass plate is developed, he hands it to Frau Roentgen. She stares, speechless, at the picture of the bones of her hand—her two rings showing clear on the third finger. She is the first person to see a photograph of her skeleton—and she has to remain silent on what she has seen for more than a month while Professor Roentgen continues to work in secret.

He finds this same invisible radiation affects gas so that it will conduct electricity. He wonders whether it will cause uranium to glow in the dark—and it does! On December 28, 1895, Wilhelm Roentgen presents his written paper (a "preliminary communication") to the president of the Würzburg Physical-Medical Society.

By that time, he had more to add: glass plates containing lead were more absorbing of the rays than those without; he had used gold, copper, silver, and platinum, also powdered materials, to see if the invisible rays were deflected, with no result. "Since I can give no explanation of these mysterious emanations," he said, "I propose to call them X rays."

His paper was published and rushed into translation for Paris, London, and Rome. The medical profession seized on the finding for aid in surgery, setting bones, extracting bullets. But not for another 17 years would it be clear that the Roentgen rays were from the same source as normal light—and differed only in that their wavelengths were a thousand times shorter.

PARIS—
January 8, 1896

Henri Becquerel had come to the high point of his career. All that he had done had led to the moment when the report of Roentgen's discoveries at Würzburg came into his hands. He had a long involvement with fluorescence, with emanations, and it was inevitable he should ask whether there were substances that also emitted unseen radiation.

Henri Becquerel was at the peak of his powers. Following his father's footsteps, he was a professor of physics, tutor at the Sorbonne, and chief scientist at the National Museum.

He was now 44, a dignified man of medium height, his full beard, touched with white, parted to fall on either side of a high, stiff, white collar, his moustache full and flowing, his eyes bright and clear. He was on the verge of an astounding discovery that would confound established concepts and shatter traditional principles. He was to write, "There was a time beloved in recent memory when progress through research to discovery was placid and peaceful. Then, the foundations of science seemed well and truly laid; but now all is changed."

In this room, uranium samples had held a place through all his life, and for all of his father's life. His grandfather, Antoine César Becquerel, in the 1840s, had been intrigued to hear that Klaproth's new element was, indeed, a true metal. His grandfather had told his father, Edmund, "I will never be satisfied with explanations they give why some chemicals and minerals shine in the dark. Fluorescence is a deep mystery and nature will not give up the secret easily."

At the age of 13, Henri learned of the Greeks, Leucippus and Democritus, some 500 years before Christ, who framed the word "atoms" from the Greek *atomos* (meaning that which cannot be divided further), for their idea that all matter was formed of indivisible particles.

His father told him how the idea of atoms had been discredited and neglected for 2,000 years until revived by European thinkers; and Edmund had often quoted Isaac Newton, who explained the forces which kept the planets in their orbits around the sun: "In the beginning God formed Matter into solid, massy, hard, impenetrable, moveable Particles. . . . "

As a young man, Henri knew that many scientists still disputed the existence of atoms, although men like Dalton, in England, argued that atoms of different types formed the various elements, which meant metals, solids, liquids, gases were all formed from different atoms. He had heard of Mendeleev's Periodic Table. But, among it all, the atom existed for Henri Becquerel—as for all scientists of the day—as a "solid, massy, hard" object. Not for a second did it enter his mind that substances that glowed in the dark were spontaneously producing energy. All his life, in the family it was accepted that, somehow, light of sunshine was stored between those solid little atoms and released slowly, gradually, by some mechanism which research one day would reveal. That was grandfather's "deep mystery."

He could remember how, years ago about the time Mendeleev's table had appeared, his father asked Henri, who had attended the Paris Polytechnic at the age of 19, to prepare some small, copper plates coated with a compound of salts of uranium and potassium. The mixture gave off a distinct glow in the dark, and Edmund wanted to test the emanation on the leaves of green plants and vegetables, and to try its reaction to charges of electricity. Would these plates *still* retain the glow, and did they throw off what Roentgen called "X rays"?

It was near the end of the month before he traced the plates; they had been lent to a university researcher named Lippman. Before they were returned to the museum, Henri was planning his experimental campaign. It would be fashionable for the envious in the next few years to describe his results as "accidental"—a cruel, unjust judgment. Typical of his training, of his family's method of approach, he was systematic, and progressive. It was the last peaceful and unhampered work he was to accomplish.

Roentgen's invisible rays had affected photographic plates. Did the same rays stream from uranium salts? If so, would they, too, have the same ability to affect photographic plates? And how could he do the test and make it foolproof?

In February, he started his series of experiments. He laid one of the two uranium-coated plates in a drawer with an unwrapped photographic plate and left it there overnight. When he developed the photographic plate, he found a fuzzy, indistinct exposure which gave no sharp lines. A clearer definition was required to eliminate the question of whether some kind of gas was given off which affected the chemicals of the photographic plate.

He cut a similar-sized copper plate and drilled holes through it and laid this over an unwrapped, unexposed photo-plate with the uranium sample, facing and on top, and again left them in a drawer. This time the exposures were sharp, showing where the holes had been drilled in the surface plate; elsewhere, the photo-plate was clear.

Now he planned more exhaustive tests. There was doubt that these emanations were X rays; they still could be stored in sunlight. He put the samples in a dark cupboard for days, excluding all light. Again, he placed the uranium in contact with the holed plate and the photo-plate. He left them there while he went on to test the effect of uranium's invisible rays on gas. By February 20, 1896, Henri Becquerel was writing a report to the French Academy of Sciences: "There is an emission of rays without apparent cause. The sun has been excluded."

He found the same invisible rays would ionize gas—change the gas atoms so they would carry a current of electricity—a finding that was forerunner to the neon tube. He announced as a fundamental fact that uranium's invisible rays were Roentgen's X rays—and he was wrong. He had uncovered a more powerful, phenomenon. For some time his newfound emissions were known as "Becquerel rays," until one of his keenest students at the Sorbonne probed their nature, for a doctoral thesis, and renamed the phenomenon *radioactivity*. The student was a Polish girl named Maria Slodowska. Soon, she was to marry Pierre Curie.

PUNGAREHU—
1895

In the same year as the Roentgen discovery of X rays, at Pungerehu in the Taranaki Province of New Zealand's north island, a tall deep-chested young man was digging the basaltic soil of his father's small farm. At 24, after five years at university on scholarships and grants, he has no apparent future. He is one of twelve children who have grown up in this hand-built, split-log farmhouse. Ernest Rutherford works at feeding the wild flax into the hand-cut wooden watermill. He has a part-time teaching job that pays a few shillings; other than that, he has no money.

He had failed to win the one annual scholarship to study at Cambridge, with his self-made device for the transmission and detection of wireless waves.

With coils and wires, and two old brass knobs from a discarded bedstead, he has transmitted wireless waves to a distance of a half-mile, and has devised an instrument to detect them. Henrich Hertz, in Europe, found such waves in 1888, but Rutherford has advanced the study and the technique so that he can send and receive such waves through the thickest walls. He is eight years ahead of Marconi's first sending of the letter *S* in Morse code across the Atlantic Ocean. Rutherford's work is far-reaching in application and potential, yet the scholarship was given to another young man for his scheme for refining gold.

Then his mother comes to tell him: "Ernie, Ernie—you've won it after all! It's alright now, dear; they've given you the scholarship!" The winner had decided to stay at home and marry, and so the award had gone to the runner-up, Ernest Rutherford.

"Mother," he smiles down at her with shining blue eyes—"I promise you—those are the last potatoes I shall ever dig!"

His parents raise a loan on the farm to pay his sea passage, from Christchurch to Sydney by sailing ship, then, steamer to England. In a few weeks he is gone from the farm; he will return, briefly, to wed Mary Newton, daughter of his university landlady. He leaves for Cambridge in the year Roentgen discovers X rays.

He carried with him his brass-knob radio device. He did not dream his work on wireless waves would be put aside within a year, that he would set out on an astonishing journey into the unknown—exploration of the universe within the atom—and make discoveries that will win him scientific immortality and the final honor of burial in London's Westminster Abbey.

Professor J. J. Thomson was patient, persistent, and perceptive. He was also very kind. In the Cavendish Laboratories, in Free School Lane, Cambridge, he was master and friend to students, and supreme intellect in an institution already famed for the quality of its research.

When the energetic young colonial, with the twanging accent and voice thick from a dose of influenza, reached him, Thomson and his wife took him under their wing. One of the first students to gain a free research berth at Cavendish, Rutherford faced problems among snobbish young gentlemen paying their way. Thomson knew this, and made the path easier.

Rutherford was allotted a corner of the cellar for his radio experiments; within two months he was courted by people beyond Cambridge, and invited to demonstrate his transmitter and detector to the elite of the Royal Society in London. It was said he would be the world leader, that he would make a fortune.

However, his path was changed suddenly. Thomson, working with a cathode tube similar to Roentgen's, was chasing the quarry of the first atomic particle. He asked for Rutherford's assistance, and, soon after Becquerel's paper was read amid excitement, Thomson handed Rutherford his first sample of uranium.

"Not such an enormous jump, my boy," he said, smilingly. "You're only moving from one wave to another."

At the Cavendish, they broke the first barriers to atomic ignorance. Thomson pinned down his elusive, unbelievably tiny particle, and, still not knowing he had found the basic unit of electric current, named it "electron," after the Greek for amber, *elektron.*

Rutherford had a corner of the room, where his maze of wires, magnetized needles, coils and screens covered a big wooden table. He sat for hours on a wooden chair, now king of his domain, narrowing down the options allowed his research by the simple apparatus he could employ.

Working with the cathode tube, Thomson made separate advances; he established the electron to be an electrically charged particle and worked out the mechanism by which radioactivity and X rays modified gas so that it could conduct an electric current.

"The energy from the rays imposes an electric charge on the atoms in the gas," he explained. "I have called such atoms 'ions'; so that, when the atoms of a gas become conductive to electricity, they are 'ionized.'"

Rutherford loved the word 'ion'"; it smacked of exciting Greek voyages of discovery. He stored it away in his mind.

Then his own studies produced the first two atomic particles to be found in uranium radiation. In the tiny beam from his uranium Rutherford discovered a heavy and a light particle. The heavy particle—later shown to be an atom of the gas helium—could be stopped by a sheet of thick paper. He named it *alpha*, following Thomson's Greek allusion. The second particle was much lighter and more penetrating, and this he named *beta*. Later, this lighter particle was shown to be no more than an electron, but to distinguish it from electrons in other situations the name *beta* was retained. Only the lack of refined apparatus prevented him from finding a more powerful radiation, gamma rays. With shorter wavelengths than X rays, and thus more energetic, the gamma radiation from uranium escaped detection until a year later, when it was revealed in France.

The Cavendish epic opened at the same time as Marie Curie, in Paris, began her chemical investigation of uranium-bearing pitchblende ores—from the same Johanngeorgenstadt and Joachimsthal mines which had supplied Klaproth with his basic materials. After more than a century, uranium was being induced to yield the first details of its astonishing secrets. Marie Curie's uranium ores were to be boiled and fractionated to isolate their by-products, all discovered by the emanations which, for the first time, she named "radioactivity."

The uranium on Rutherford's crowded table was in a small lead capsule, into the side of which he had bored a hole. From this tiny aperture the invisible radioactivity streamed, as straight as a beam of light, to strike the various screens, targets, and instruments he used to reveal the components thrown off by the uranium sample.

The discoveries at the Cavendish hinted at the crumbling of long-held ideas; but, coming after the excitement of Roentgen's X rays, and the exposure of that discovery by the popular press, they made hardly a ripple in public affairs. Yet Rutherford blossomed into a celebrity; the first, small circle formed about him. He was still young enough at 27 to be boisterous and enjoy life with his coterie of friends, among them a tall, serious French student who was destined to escape the clutches of the Gestapo in World War Two—a future leader of French science, Paul Langevin.

The radiation work, the existence of alpha and beta particles, the discovery of the electron won academic attention; offers of jobs came to Rutherford. One came from his homeland, where Mary Newton waited patiently for his return. The salary was good, but he rejected the offer. He accepted the chair of physics at McGill University, in Montreal, and wrote home, "I know the pay is not so good, but the equipment there is excellent . . . and they want me to form a research school to knock the shine off the Yankees."

It must have been a joke, because the teaching of physics in America had been a neglected subject; the American Physical Society was not formed until a year after Rutherford arrived at the MacDonald Physics Building.

Montreal made changes in Rutherford; and the changes he brought to the university were never to be forgotten. He threw himself into the creation of a laboratory with little financial backing; his own wages were $12 a week. From his first supply of thorium, then just found to be radioactive, he opened his classical studies.

In 1900 he sailed home to wed Mary and returned to McGill University to welcome a new recruit to his laboratory from Oxford, Dr. Frederick Soddy.

Now he was at the first rung of an astonishing climb in the history of physical science; for almost four decades he would bestride the scene of experimental physics.

J. G. Crowther would say of him, "He conducted research like a Homeric hero . . . brandishing the sword of discovery as he penetrated the unknown—at the head of his band of youthful followers."

MONTREAL—
April, 1902

Dr. Soddy bent to enter the solid wood door, ready to pick his way over the cables on the dark floor—and stopped in surprise. The overhead bulbs glared naked into the basement laboratory; even more unusual, Professor Rutherford was not squatting in the gloom, peering down his microscope or fiddling with his coils of wires and magnetic devices. He was on his feet, white coat flapping around him, prowling up and down the open space.

Catching sight of Soddy, he broke his cardinal rule that peace and quiet should reign in his laboratory and let out a great shout. His blue eyes glinted in the light of the bare, electric bulbs; his hair seemed to bristle. Flinging his arms wide, he leaped into the air, came down with his feet apart, knees wide, his body and arms swaying; and then he broke into a strange, bellowing chant.

Nurtured in the sedate Victorian environment of Oxford, where senior academics were models of decorum, 23-year-old Soddy, doctor of chemistry, was dumbfounded at this antic. Never before had he witnessed a Maori *haka.*

Rutherford's voice boomed through the cellar: "We've done it Soddy! We've got it! Proof, my lad—incontrovertible proof. All we've got to do now is write it up. It really is astounding—it'll knock them over. They'll quiver and argue . . . I know! But it's all come together and it fits. It's big, my lad, big! And I thank you for your chemistry which completed our picture."

They had talked of the potential in the work; Soddy knew the breakthrough was historic. It wasn't like most science, evolutionary; it was totally new ground. In the gloom of the cellar, sitting, waiting for their eyes to get used to the dark, they had debated and cogitated for many hours on the meaning of what they saw through the microscope—tiny scintillas of blue-green light where Rutherford's radioactive beam struck the target atoms in a sheet of thin foil. Collisions! Yet what was the origin of these bits of matter, these particles spitting from his sample, like a stream of bullets from a Gatling gun, smashing into the foil sheet and showing tiny sparks of released energy?

Rutherford now had the answer; and he was ecstatic:

Soddy! Those atoms are exploding! They are bursting open, flinging off bits and pieces of their structure—constantly and without let-up. But that's not all! If you ponder the old laws of conservation, then you have to ask: what happens to atoms which throw off these parts of their structure and composition? Yes—you have to admit it—they *must* change! So, they *are* destructible, Soddy! They *are* changing—they are throwing off energy and changing from one element to another. And you know what that is, don't you? It's transmutation, my boy! Natural transmutation. But we dare not say that in our paper. They'll call us alchemists, charlatans, and try to cut off our heads. No, we can't say that. We can't tell the gray-beards that atoms explode all on their own, without human aid. We'll dress it up a bit. Let 'em get used to the idea that nature changes herself. We'll call it "spontaneous disintegration." It sounds more respectable that way.

Rumors of the discovery ran through the university like a bushfire; and Rutherford's achievement was at once clouded with disbelief.

Through the ages, men of learning had dreamed of supreme power from the legendary philosopher's stone, which would endow them with the ability to transmute one element to another—to change base metals to gold and silver; and here was the ebullient Professor Rutherford—in the face of all modern science and the known laws of nature—claiming this as taking place in his laboratory! It was not just an advance of knowledge that was reported, it was a departure from known principles. It was revolution. It was repudiation of accepted notions of the structure of all matter . . . that atoms were hard, solid, like billiard balls, and absolutely indivisible. They might be different in arrangement, and size, and color, and weight, but they *were* indestructible. Everyone accepted that—except Rutherford and Soddy!

The seniors of the faculty formed a small deputation; they came to the cellar to see Rutherford. "Please, Ernest," their spokesman urged, "don't publish without reconsidering and re-checking your results once again. Don't publish your ideas on this thing, not yet. We're afraid you might bring damaging ridicule on this university."

Rutherford's laughter rolled through the cellar, trumpeting his certainty.

"Do you think this is a flight of wild fancy?" he roared. "Whether we like it or not—whether the world likes it or not—it happens in nature. Atoms throw off energy and bits of their structure and change from one element to another. And that is transmutation. So—what does uranium change into after it explodes? I can't tell you yet, but I'm going to find out. I'm after the pattern behind all this orderly destruction."

Rutherford enjoyed confrontation of tradition. He had not had such fun since leaving Thomson in Cambridge four years ago—apart from his few months in New Zealand and the delight of marriage and the birth of his daughter, an only child.

His world revolved around the atom, and when told he had a daughter, he decided on her name. "Mary," he said, "when Thomson told me about the effect of electricity in gas—turning the atoms into what he called 'ions'—I loved the word. We'll call her—Iona!" Mary's conservatism defeated him from the start. Rutherford's lovely Iona became Eileen.

The events of the outside world that made impact on his mind during these years were Marconi signalling the letter *S* across the Atlantic and the discoveries of Marie and Pierre Curie, who were working along the same path, seeking to solve the mysteries in the nature of uranium. Their trail too, had commenced with Becquerel's emanations from a uranium sample. The Curies had chemically broken down their samples to show two entirely new elements in uranium ores—radium and polonium. Entirely puzzling to their minds, these two substances were thousands of times more radioactive, more energetic than uranium itself!

Rutherford's work was as widely reported. In the following months, he was elected—at 32—to membership in the Royal Society and invited to London to accept membership with an address to the assembled Royal Fellows; and, by another post, came his invitation to visit Paris, as an honored scientist, as a friend of Paul Langevin, associate professor at the Sorbonne, and to meet and talk with the Curies.

PARIS—
August, 1903.

A blanket of torpid air hung over Paris, and in the Langevin's roof garden the noise of the city was smothered. They were uncomfortable, standing, facing Pierre Curie; the women sweating in tight bodices under long evening gowns, the men brave in stiff shirts, high collars, and dinner jackets. All waiting. Rutherford was in the center of the line, with Mary on his arm and Langevin and his wife at the other side. Next to Mary was Marie Curie and, with her, the French chemist Jean Perrin. Pierre Curie stood with fingers resting in a waistcoat pocket, watching, waiting until their shifting, restless movements stilled. With a quick movement he took from his pocket a small glass tube, not two centimeters long, with sealed ends and coated with zinc sulphide. Pinning this between thumb and forefinger, he held his hand above the shrubbery, against the night sky.

At once, the heat, the discomfort, were forgotten. They gazed in wonder at the sight of a man with starfire at his fingertips—but much brighter than the Milky Way: A totally different kind of starshine gleamed from Pierre Curie's tube, a mysterious cosmic blue-green glare that flooded the garden with light and filled their faces with shining amazement.

For Ernest Rutherford, the vision in the Paris garden was immense—the empirical validation of his insight. Here, in full flood, were those tiny blue sparks he'd detected in the cellar at Montreal.

Through his sleeve Mary Rutherford felt the tension of the surging excitement in a scene she would remember all her life. And not only Ernest seemed transfixed; by her side, Marie Curie gazed on the radiance, bewitched, as though she had not seen it before. The world had been told of the Curie saga, of the hunt and discovery of radium and its sister substance which Marie had named polonium, after her native land.

Mary Rutherford knew of the epic of slavery in that old cadaver shed in Paris, of the endless boiling down, the chemical reductions and precipitations of the ton of uranium-bearing ores brought from those same deposits which, a century before, had revealed uranium to Klaproth. In many ways, Marie Curie had followed the German apothecary's patient chemical pathways to the revelations—to radium, radioactivity, to polonium, and to discovering emanations from thorium.

Like Ernest Rutherford, she and her husband, Pierre, had opened a new road to knowledge.

Then the magic moment was done. Pierre Curie lowered his arm and put the glass tube back into his waistcoat pocket. They filed into the hot dining room, cloudy with candle smoke, tongues loosened in animated talk.

At the table, Mary Rutherford listened. Ernest and Pierre Curie dominated the discussion; Marie sat quiet, her eyes glowing as her husband talked. He said Marie had faced two explanations for the radioactivity that Becquerel had discovered in uranium. The first was that uranium borrowed its radiation from some unknown exterior source. It was not as absurd as it sounded; it was known that certain invisible penetrations traversed space; there might be others. The second possibility was more challenging and had been given validity by the work of Dr. Rutherford. It was that the atoms drew this energy from *within* their own structures. In a public statement, Pierre Curie was to enlarge the point:

"Once we accept this theory, many perplexing things become understandable. You see, it means a veritable change, or transmutation; but not as the alchemists saw it—rather, it is evolution among the elements, as it has been shown by Darwin to occur among living species. This may sound strange at first, yet when you ask: why do we always find Marie's radium and polonium and thorium in association with uranium? It emerges as the reason! We believe, now, that stemming from uranium is a whole family of radioactive elements. And, for some reason, these substances, polonium and radium, can be a million times more energetic, or radioactive, than the parent uranium. The importance for physics is evident. The consequences will have an impact on all branches of science."

In the Langevin's dining room Mary Rutherford saw her husband gazing at Pierre Curie's hands, and she looked at Marie's hands; like Pierre's they too were scarred with brown tissue and the fingers were dark and wrinkled at the tips—like old prunes.

She asked them about this energy, this light that came from radium. Was it heat, like hot water? Was it fire, like the sun, or burning wood?

Rutherford took the point, as recorded in his papers "We can answer that quite simply—we just don't know. It's new, it's perplexing, but we are on the fringe of understanding and we're going to find out.

"We know there appears to be no reduction in the energy coming from radium—but that doesn't mean it can go on forever; it might go on throwing out radiation for a very, very long time— but we just cannot see the source being inexhaustible. We have to admit, however, we have made a discovery of forces and power beyond present knowledge, quite beyond imagination. It is a revolution; it will upset many concepts in the established order of our environment . . . we are walking into strange territory, a no-man's-land of scientific mystery. I think we are going to turn up many more strange and wonderful marvels of nature. It is a thrilling point in the struggle for knowledge. . . . God alone knows what we shall find. . . ."

Rutherford was to remember Pierre Curie's intense serious-ness: "I am anxious, concerned, at what we are doing! You see, radium and such substances could be dangerous in criminal hands. We should ask—will mankind benefit from knowing the secrets of nature? You know, it has to be handled carefully—it burns my skin through my clothes; Professor Becquerel com-plained to Marie of blisters on his chest when he carried it to his rooms. I am fearful that in stronger concentrations it could lead to paralysis, or death. I ask, what if such a dangerous force falls into the hands of warring men, if it is used by the criminals who lead the peoples towards war? How shall we justify what we do?"

The Rutherfords left for their hotel soon after. As they took their leave, Rutherford's sympathy went out to Pierre in his inner struggle. He had the feeling Pierre was ridden by a sense of disaster. Neither knew the accuracy of their premonitions.

They were never to meet again. Pierre Curie uttered the same anxieties in Stockholm, four months later, when the third Nobel Prize for Physics was awarded to him and his wife, and to Henri Becquerel. Then he drew his parallel between radioactive ele-ments and Nobel's discovery of high explosives. "They have en-abled men to perform admirable works," he said. "They are also used as a terrible means of destruction."

Less than three years later Pierre Curie's head was crushed into a Paris gutter by the wheels of a brewer's dray.

Rutherford left Europe with that garden scene burning in his mind. It had given strength to his resolve to state a new age for the earth when he had lectured to members of the British Institu-tion, despite old Lord Kelvin, crusty doyen of British science, who was waiting to pounce. Only the glow in the garden in Paris, the incredibly slow rate of decay in uranium and radium, had given him the courage and the resolve.

They were yardsticks of time; one could now measure the years they had been casting off energy, and, so, they were na-ture's clocks. These radioactive clocks now made it certain that the world was at least 700 million years old—and Lord Kelvin was sitting up. The Kelvin age for the earth of 100 million years had been made, he recalled, with the proviso that a new source of energy would have to be found to change his arithmetic—and it had. That new source was radium.

Now it was back to McGill and the basement in the Mac-Donald Physics Building. Soddy was to return to England, where he should try to find chemical proof that uranium gave birth to radium, while he, in Montreal, would follow the process of "spontaneous disintegration." He would miss Soddy; he liked his work; but there were other bright young men keen to work on radium.

LONDON—
1905.

Young Robert Rich Sharp was a true product of the English upper middle class. Charterhouse and Christ Church, husky, a champion cross-country runner, he was just down from Oxford, with nowhere to go. His scholastic ability was too low to win a place in the Indian Civil Service; he had toyed with becoming a schoolteacher and applying for a post in the newly formed Soudan Service. Then, out of the blue, mother had pulled strings and he had an interview with Earl Grey—about to take up his appointment as Governor General of Canada—who sent him to his younger brother, George Grey, a legendary figure in Africa. George Grey had taken the first expedition from Bulawayo to the Congo, and had discovered copper outcrops that promised great potential; he had raised Grey's Scouts, from tough pioneer-prospectors, to fight in the Matabele Rebellion, and his daring and dash had made him famous.

In the Clements Lane office of Tanganyika Concessions Ltd., Grey eyed his applicant, in smart lounge suit and tie, visualizing him in the bush:

"My brother tells me you can run ten miles without halting?" That was so; Sharp had won the London Athletic Club's Challenge Cup. "Can't help feeling that a chap who can run ten miles and beat all-comers must have some good in him." He had a pronounced stammer.

Did Sharp know the rough life, the dangers of trudging the

wild land, the discomfort, far from home and family? It roused only excitement, a challenge.

"Well," stammered George Grey, "we'll give you a chance—ten pounds a month with your rations. But you'll buy your own gear, and weapons. You'll have three months or so to prepare. We'll arrange lessons on rock identification with an old friend of ours in Regent Street; and you'll be given a grounding in survey work, using a sextant, star-sighting and map-making and the like, at the Royal Geographical Society. The rest is up to you. Happy to have you join us. We'll let you know about your travel arrangements."

Grey's face was a sallow tan from long exposure to the African sun and the illness the jungle inflicts; yet, to Sharp, he seemed to glow that day as he opened the door to a life in primitive Africa. There, Sharp was to grow into one of the great Bwanas of the day; to fight in the coming World War I and win exceptional decoration; and in Africa he was to find the mine that would help to end World War II.

FRANKFURT AM MAIN—
September, 1905.

Herr Heinrich Hahn ran his family on good Prussian lines. Running his business, making picture frames and gilding mirrors in the basement, selling them from the ground floor shop, he set a firm example of diligence, industry, and quiet living. His son Otto, who should have been training to be an architect, was fiddling with chemicals and being paid nothing, even though he had obtained his doctorate. For two whole years this had gone on—and for a year before that, when he was twenty, Otto had served as a volunteer soldier with the 81st Regiment, emerging as vice-sergeant.

At 23—still with no job—Otto was asking for more money so he could study in England! Of course, there was a chance he would get a lifetime post out of it; but who could be sure? Otto had told him this important man, Director Fischer, father of the famed Nobel Chemistry Prize-winner, was looking for a promising young chemist for his company—the Kalle Organization. As recorded in Hahn's autobiography:

"Professor Zincke has recommended me, Father! He thinks I

should go first to London—for at least six months. If you agree, he will arrange a place for me to work and to study the latest trends . . . but it is up to you because there can be no salary until I can take up the post here."

Herr Heinrich Hahn gave his third son the chance and sent him on his first of many journeys from Germany—off to London, where he was to work with Sir William Ramsay in the Chemical Institute, there to be confronted with the residues of a quarter-ton of ores, from Ceylon, in which was a substance that would have bearing on his future career.

SWITZERLAND—
1905.

While the groups in Montreal, Paris, and Cambridge worked in dingy cellar laboratories, in the Government Patents Office in Berne, working after office hours over a wooden-top desk, a 26-year-old German-born Jew named Albert Einstein accomplished one of the great feats of human intelligence. He was then unknown, a "stranger with no university credits, no university chair, no claim to such audacity."

His creation was of the mind, and he gave a fourth dimension to the universe. He did not build the atomic age, nor did he make material discovery; he gave it form and meaning. He gave explanation to the radium glow that lit Marie Curie's eyes, to the blue-green scintillas on the sheet of foil in the basement at McGill, in Montreal. His first paper, in 1905, was titled "On the Electrodynamics of Moving Bodies." It was the first of the writing that was to become part of the general theory of relativity, and in those first pages was opened the stunning new concept of the nature of matter as being a form of energy locked into structures in a "frozen" state. The analogy was that energy could be fluid—light, heat, electricity—or fixed into a firm condition, like water turned to ice.

This far-reaching proposition said all matter was energy, that all energy was matter released.

There were many issues in the work Einstein produced, of time and space, and the weight of moving bodies; but it was his famous equation, $E = mc^2$, that gave explanation to the energy streaming from uranium, and its family of radioactive elements.

This equation showed enormous forces captured in the structures of the atom; the heavier the atom, the greater the contained energy. The E in the equation signified the energy equivalent (expressed in ergs), while the m was the mass, which was meant to be multiplied by the square of the speed of light (c^2). It made a mind-boggling total. It predicted the energy release from a fragment of metal could be great enough to devastate a city; that the energy in an ounce of matter could turn a million tons of water to steam. In 1905, it was too much for most people to comprehend; fable has it that less than a dozen men in the world understood what Einstein was stating. Over the years it dawned that he had demolished the age-old law of conservation of mass by showing the equivalence of energy with matter, by indicating that energy bound itself into the particles that formed atoms.

Einstein threw a bright light on what was happening in radioactivity; the understanding of how it took place was left to the ceaseless probings of the experimental physicists. A Polish scientist exclaimed; "A new Copernicus has been born!" It was years, however, before Einstein's remarkable feat was widely understood.

MONTREAL— 1906.

When Ernest Rutherford first set eyes on Otto Hahn, in the physics building in Montreal, he thought he looked a miniature version of Kaiser Wilhelm. Taut, stiff-backed, close-cropped hair, an upswept, pointed moustache, his accent and his manner were still very Prussian and had mellowed only slightly during his months under Ramsay's tuition.

For Otto Hahn, Rutherford was a surprise: a man of ease, of humor. As he says in his book, he found Rutherford to be very direct, and blunt in speech.

"I've heard you make claims for discovering a new element, my boy! Ramsay's a dear fellow, doing excellent work on gases; but radioactivity ... ?" Head tilted back, eyes half-closed. "We've looked at it briefly here—can't find anything new. And I should read you this," fishing a paper from his jacket pocket.

"My friend Boltwood of Yale says 'Hahn's new element looks like a compound—of known thorium and stupidity.' What've you got to say to that?"

Hahn had deliberately sought his place in Montreal; Rutherford's publications had fixed his laboratory as a place of great potential, from which came the claim that energy in the atoms of uranium and other radioactive substances, must be "at least 20,000 times, and maybe a million times, as great" as was released in any chemical change. Otto Hahn had come to work, to research, and, again, without pay.

He knew he was under test with Rutherford's challenging question. He stiffened.

"I hope I am able to convince you. I happen to have with me the sample of radiothorium which I extracted from the ore from Ceylon."

Ramsay had presented him with the residues of the quarter-ton of radioactive ore. It was precious then, for there were few known deposits of radioactive material—a small pocket here and there and chemical supply firms seeing not enough interest in it to make marketing worthwhile. So it was treated carefully in London. When it came to Hahn it had been examined in various ways, but he precipitated and reduced the matter with heat and acids and had found unexpected levels of activity. Through crystallization, he came to this little example, some forty thousandths of a gram of an element that was 250,000 times more radioactive than thorium itself. He had found lead, he had found uranium, he had found thorium, and he had also found this tiny sample of a new matter which he named radiothorium. He said, "I will be happy to prove it different from all other elements known."

It was a case of mistaken identity that was to be repeated again and again.

Otto Hahn's first work for Rutherford was to produce a scale showing how alpha particles from different radioactive samples traveled differing distances, even though they came from similar chemical elements. There seemed nothing startling about the results, at the time. It was another of the many projects which Rutherford set in motion in his laboratory where, starved for funds, they spent many hours improvising, soldering coffee and tobacco tins together, using cheap tinfoil and, sometimes, old pieces of wire and string, simple magnets, and primitive electrical gear; they also had to pioneer methods of investigation. At that time, the expense of $100 would be a carefully assessed matter in the laboratory where they were laying the base for the vast complexity and technological sophistication of the future generations of nuclear power stations.

Hahn's alpha-particle scale, however, was the first solid hint of order behind increasing confusion caused by new finds of so-called radioactive elements, such as his own radiothorium. One after another, these new "elements" were announced until they were a jumble of perplexing radioactive substances that would not fit into the Mendeleev table. Mystery and perplexity dogged the work, until the "isotope" solution arose—fittingly, from Soddy, who was working with Ramsay, more or less having changed place with Hahn.

Soddy reasoned that many of these seemingly different radioactive "elements" were chemically identical with the substances from which they were extracted; that they differed slightly only in atomic weight, or mass. It meant that atoms in the same place in the same chemical structure had cousins, atoms closely related, but different somehow in their structures. He called them "isotopes" (from *isotopic,* meaning in the same place).

Soddy's finding looked to be a simple step forward into Rutherford's no-man's-land. Not for decades would its full impact burst into public consciousness—when it led to the incineration of Hiroshima, to the first atomic reactors and the great nuclear enrichment industry that arose in secret during World War II.

However, from the MacDonald Physics Building at McGill, Rutherford wrote, "There seems to be reason to suppose that the atomic energy of all elements is of a similar order of magnitude. If it were ever found possible to control, at will, the rate of disintegration of the radioactive elements, an enormous amount of energy could be obtained from a small quantity of matter."

BERLIN—
Spring, 1907.

At the Chemical Institute in the University of Berlin, Dr. Otto Hahn made his first important academic advance; he was appointed *privatdozent*—assistant professor—to the head of the chemistry school, Professor Emil Fischer. Again it was not a paid position, but it was the proving step to full professorship; and Hahn worked hard in his downstairs "wood-room," a former carpenter's shop.

His task of seeking a "mother" substance from which radiothorium sprang was regarded as outlandish by run-of-the-mill students, and his appointment earned hostility and resentment. "Amazing what gets to be made assistant professor these days," was a remark he overheard; and his presence at chemistry symposia and discussions was received without warmth. However, this was not so with physicists working in the university, and Hahn made a habit of walking to the classes conducted by Dr. Heinrich Rubens.

With summer gone in isolating what he took to be the "element" that gave birth to radium—still not knowing Soddy's isotope theory—and in learning that his old critic from Yale had also retrieved the substance and named it "ionium"—Hahn walked over to a physics lecture on the afternoon of September 18th and found a young Jewish woman in the audience. Women in science were rare enough in 1907, and she was therefore of great interest to all the male students, especially as she was quietly appealing. Otto Hahn was attracted to her reserve and her intelligent eyes; her solitary manner fitted his own isolation. They talked after the lecture.

Her name was Lise Meitner; she had won her doctorate in Vienna on the conduction of heat and had been introduced to radium research. They had a common ground at once; she told Hahn she was planning to do theoretical physics under the star professor, Max Planck. She had already written two papers on alpha rays. Since the Planck lectures did not keep her fully occupied, she asked Hahn's professor if she could work with Hahn in the "wood-room." This presented Fischer with a problem. The rigid rule was that the Institute did not accept women in his faculty. Hahn pleaded and won a concession: Lise Meitner could assist Hahn in the radium work "provided she did not enter other laboratories where male students were working."

Happily, they began their association in October and opened a famous partnership that would last 30 years and end with Hitler's racial laws.

MANCHESTER—
August, 1911.

Mary Rutherford played the piano while the atomic nucleus was discovered. Her guests had waited for more than an hour for high tea, but Ernest was still at his laboratory, late on this Sunday afternoon.

The weekly entertainment of senior students, staff, and colleagues went ahead. She chatted with them for a time, and they talked among themselves; but she was so restless, one of them asked her to play the piano. She was midway through her second piece when she heard his heavy tread on the path and his voice booming into the house with excitement.

There was no thought, nor time, for apology in the exhilaration of this occasion. Rutherford strode to the center of the room, delighted to have a perceptive audience on hand. He stood with his coat swept back, hands on his hips, head tilted back, hair ruffled, eyes aglow with triumph.

"I've got it!" he trumpeted. "At last—at long last—I know what the atom looks like! I've found out how it is structured."

Among his guests was Charles Galton Darwin, grandson of Charles Darwin, the founder of evolutionary theory; and he was caught up in the wonder of the moment. "I count it among the great occurrences of my life," he wrote. "I was actually present within half an hour of the birth of the planetary atom."

The great leap came thirteen years after Rutherford opened his hunt for the first flying particles—the alpha and beta fragments in the radioactivity from uranium. He had arrrived at this memorable conclusion a decade—to the month—after seeing the blue-green glow of radium acting on the coating of zinc sulphide screen in the Paris garden. That same light, coming in tiny flashes on a zinc screen, had led him to the inner secret of atomic structure.

Years of persistence aided the insight of genius. In 1906, when Hahn went back to Berlin, Rutherford rejected lucrative offers from America to work instead at Manchester because the laboratory had been equipped with the latest instrumentation. There, also, he inherited a young German student of special gifts who was to help fashion methods of detection of radiated particles by mechanical means. His name was Hans Geiger, and his counter was to provide the tool to detect the flow of atomic particles.

In 1908, Rutherford was awarded a Nobel Prize for chemistry, which indicated the uncertainty still shrouding the field of atomic research. Together, Rutherford and Geiger would wait for hours for their eyes to become accustomed to the dark so they could look through a magnifying glass and see where alpha particles collided with atoms in the target foil—to give off the infinitely small scintillas of light, then to deflect on to another sheet of foil, to cause a second tiny flash.

Rutherford used a striking allegory in explanation to his guests: "We found that many radiated particles are deflected at staggering angles—some recoil back along the same path they have come. And, considering the enormous energy of the alpha particles, it is like firing a fifteen-inch shell at a sheet of paper and having it thrown back to you!"

Some astonishing force was responsible for repelling his atomic shells; he came to the final, historic conclusion: the alpha fragments had a speed—deduced from his magnetic experiments—of about ten thousand miles *per second;* yet, some awesome energy could fling them aside, or push them back, at less than a 90-degree angle. The picture of the recoil took shape in his mind; the atom had a small, central core—later to be called the nucleus—and this was so heavily electrically charged that it repelled the high-energy alpha particles. He reasoned that this atomic core was infinitely small, judged by the deflections, and he first put it to be only one-ten-thousandth of the total volume of the atom. (Later, it was shown to be nearer one billionth.) It was possible, now, for him to vision an atom, magnified one thousand million times, to the size of a football, with the core, or nucleus, of the size of a dot. And—as with a football—the atom was mostly empty space. There was only this tiny, central core, surrounded at great distance by a field of whirling electrons—orbiting the center like planets. To make this possible, he argued, the central core had to be positively charged and the electrons negatively charged.

Rutherford had found the font of all nuclear and chemical energy; he had revealed the basic structure of the citadel that provides all light, heat, and the processes of life itself.

The impact of this deduction was still to come. It did not at once arouse much excitement—people found it hard to accept that all matter, including the atoms of their own bodies, was mostly space! Within the growing field of atomic physics, however, it soon made Rutherford's laboratory the mecca for young physicists; and in that same year of the discovery of the planetary atom, a young physicist from Denmark came to Manchester to work under Rutherford. His name was Henrik David Niels Bohr.

A bulky Dane with a massive head and a rubbery face with a near-squint, he was at once liked by all who knew and worked with him. His habit of speaking accented English in a kind of monotone often made it hard to follow what he was saying, but that suited the dictum Rutherford imposed on all his workers: "Speak softly—others are working."

Niels Bohr was the first of the Rutherford protégés. With Rutherford, he saw there were questions to be asked about the planetary atom—that the system of a shell of circling electrons could not ever be stable; that they would be liable to a loss of energy and would fall back into the nucleus. He revived a theory from Max Planck.

It was called the "quantum theory," and it suggested that light was related to the movement of electrons in nuclear orbit and was not emitted in a straight, continuous stream, but in separate packets—in "quanta." It was this idea, developed a quarter of a century later, that founded quantum mechanics. Bohr used this theme to explain how atomic electrons were committed to single orbits: that, if they were caused to jump to a smaller orbit, they had to shed some of their energy in the form of radiation of light of precise wavelength or color. It was the first explanation of the emission of atomic energy. It was Bohr's first notable contribution.

There were still many mysteries to be tackled in the years prior to World War I. There was now a collection of brilliant minds in Rutherford's group, among them Dr. Henry Gwyn-Jeffries Moseley. Before the war, Moseley had bombarded each of the known elements with a beam of electrons to show that the number of electric charges in each nucleus was increased by regular steps between each element in the periodic table. He was said to have "called the roll" on the elements and to have provided a valuable yardstick for the atomic scale. When war broke out, the Army claimed him, despite Rutherford's angry protests. In Gallipoli a Turkish sniper put a bullet through his brain.

KATANGA, AFRICA—
August, 1915.

Bwana Sharp sat motionless and quiet in the mushito bush, waiting for the leopard, his Rigby-Mauser .303 resting on his crossed legs. The big cat had killed a wild pig during the night, and he knew it was bound to return for breakfast.

It was a change from the tedium of the traverse across the Katanga province, searching for outcrops of possible mineral deposits. It also relieved his restlessness. Here he was, in mid-1915, his brother Charlie killed on the Somme, still trying to get back to Britain and join the army. He worked now for the successors to the original Tanganyika Concessions—the Belgian Union Minière Du Haut Katanga; good people, but they did not seem to understand he wanted a part in the fight, that he wanted some revenge. And they would not let him go while there was prospecting to be done.

He could hear the leopard in the bush about 30 feet away; its teeth were tearing at the dead pig, but Sharp could see no more than a few spots. He records in his book how George Grey had died from the mauling by a lion he had wounded; and, if he missed, this brute could cross the distance between them in a single bound. He fired—a single shot; and the leopard dropped.

Elated, he pushed past his chattering boys and strode ahead, leaving them to skin the animal and catch up to him with the trophy.

He was then west of the company's big Star of the Congo mine, near Elizabethville (to be named Lubumbashi) and some 50 miles southeast of Kolwezi. The country was mostly flat, except for the kopje, the small hill—more like a big mound—rising from the dense evergreen bush. There were copper signs among scraggy trees and that was what he was after.

There were scars at the top, indicating that natives had worked the site at some time past. It did not look like much, but it had to be pegged and listed. While he waited for his boys to catch him up, he literally kicked his heels; and the tough bush boots turned over the piles of rocks. As he scuffed them apart, the gleam of yellow caught his eyes.

Sharp picked up a hand-sized rock, laced with green streaks and ochre—and the brilliant yellow; there were larger rocks strewn across the surface, some obviously weighing several tons; and it was the weight more than the color that first captured his interest. It was heavy, very heavy, the first indication to the prospector of valuable metals. Then, his mind was reaching back: somewhere he had seen something like this. Yes! At Luisiwishi, near the Star of the Congo. It was a sample in a collection, and they had told him it was loaded with useless uranium but had a trace of radium; and this was similar, but more highly colored.

This was all float ore; yet, the hog-back of the kopje promised a chance of a reef; and that was worth a little work. He shouted to the boys to hurry, to bring the tools and instruments he needed to locate the deposit, whatever it was.

They cut cross trenches along the ridge, and there was a definite reef; it probably had lifted the kopje from the river flat in geologic times. It extended right across the kopje and had some depth, but was workable from the surface. He picked two rocks and called his head boy, Malupenga, and told him to send them to the assay office at Likasi, about 20 miles to the northeast.

He questioned Malupenga, "What people live in this district?"

"They called Bayeke, Bwana."

"Is there a town, or village, where they live?"

"That way, Bwana, half-day walk—called Shinkolobwe."

That would be the name for the deposit; if it became a mine, it might be something else. He fixed its position, wrote his notes, and went back down the slope heading west, wondering about the weight, and the colors, and whether there was really some worthwhile radium among the other things.

The Shinkolobwe uranium deposit was very old. It had lain on the kopje summit for much more than 600 million years. Defying gravity in a long series of geological steps, it had floated from inside the fermenting planet, like slag in a furnace, because its structures—the way its atoms formed into lattices—had no compatibility with the other jostling, steaming abundances of earth's inner regions. It had conformation that was too big to fit

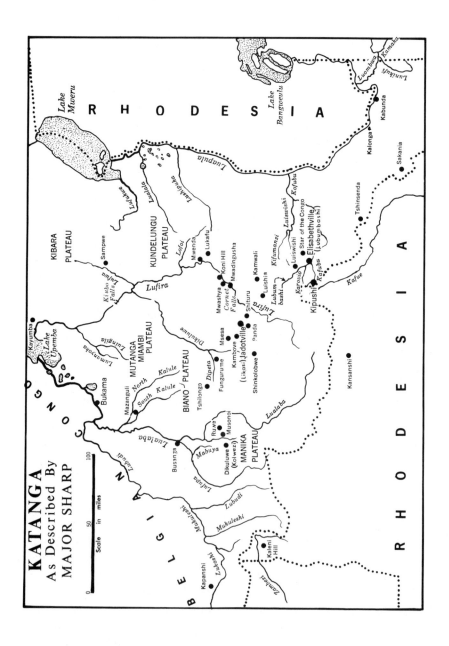

KATANGA
As Described By
MAJOR SHARP

there; it was slowly concentrated with its own kind of material—and, over the endless hundreds of millions of years, it was squeezed to the surface, in streaks and reefs, like gold. To geologists it was "a spur wedged in a fold fault."

By this process uranium had reached the surface of the planet in many places—but none quite like this outcrop near Shinkolobwe, in the very heart of Africa's 11.6 million square miles. There was no other deposit like this known in the world, nor would there be one with such a place in world history. The riches in its strata, the first bounty of radium it would produce, would pale against the opulence of uranium in ratios unparalleled. Where known deposits—including those still to be found in Canada—were workable with oxide yields of one percent or less, this mine was to average 68 percent pure uranium. And it rested on the surface! Here and there were loose rocks weighing as much as seven tons—with enough basic material in each to shatter a major metropolis, once the method was known.

A week later, Bwana Sharp returned to his Shinkolobwe kopje. The assay was promising; and, with it, came a letter from Union Minière releasing him from duty for service in France. He completed his task. He set up the pole required by law to validate the claim; it did not demarcate a rectangle as in most countries—it claimed a circle of land with a radius of 1,000 meters from the wood pole. It carried a sheet of zinc on which the word RADIUM was cut in letters six inches high. It made no mention of uranium, which had little value other than for tinting glass in ceramics.

He wrote out details of the claim, and sealed them in an envelope which was sent to the Union Miniére office at Lubumbashi; in his note he asked that the claim be lodged at the mines office of the Belgian Comité Spéciale du Katanga.

And then he went off to war.

MANCHESTER—
1919.

For two years Rutherford had been firing his alpha "shells" into light elements. His targets had been pure nitrogen and ordinary air—a mixture of nitrogen and oxygen. In both cases, all hydrogen had been cleaned from his target material. In both cases he found, after bombardment, that nitrogen had disintegrated and ejected a hydrogen nucleus.

Again, he had made a classic discovery: The core of the hydrogen atom was blasted from the core of nitrogen. Rutherford had achieved the old dream of alchemy; this was the first time man had accomplished transmutation. And the hydrogen core was a building block of all atoms, all nuclei. He was to name it the proton; and the proton was the stimulus to his remarkable insight, the keenest in this field at that time in the world. He saw in the heavy atoms, up to uranium, that other particles had to live side by side in the fields of force that held the nuclear core tightly packed, in its place, inside the shell of electrons.

The war bore heavily on him. He was deeply saddened—and offended—at the shocking waste of Moseley's talent; he took part in the scientific direction of the war effort and served on a committee which designed the ASDIC equipment for detection of submarines below the surface by means of sound waves; but he was often so engrossed in atomic research that he forgot to attend meetings. An Admiralty official rebuked him for missing a meeting, and he chided the man, "Talk softly to me, my boy. I have been engaged on work of far greater importance than a mere war!"

Yet, war and his work were linked; he remembered Pierre Curie's anxiety. What if this power should fall into the hands of warring men? In 1904 he had glimpsed a possible way atomic energy might be released; when he was investigating gamma rays he had remarked on the chance of "uncharged bodies projected with great velocity." The worry had come back to him in 1916 when he was pressured into a public lecture in the London suburb of Islington. He told his audience, "A mere pound of uranium, on the scale we see possible, might liberate the same energy as burning 100 million pounds of coal. Scientists are trying hard to find a way to release this energy at will. Personally, I hope they do not succeed until man has learned how to live with his neighbours in peace."

Unwittingly, his work was edging toward that achievement. The transmutation at Manchester, the proven existence of the proton—the hydrogen nucleus—in other atoms, gave him a base for a projection that was to come at the end of 1919, when he succeeded Sir J. J. Thomson as professor of physics at the Cavendish and went back to where it had all started.

KATANGA—
1919.

Major Robert R. Sharp, D. S. O., M. C., Croix de Guerre with palms, hero of the Royal Field Artillery, returned to the Union Miniére's office in Lubumbashi.

A shock awaited him. No action had been taken to register the Shinkolobwe mine. The demarcation pole with its six-inch-high sign, RADIUM, scratched on a zinc plate proclaimed company rights over a circle 2,000 meters wide, but it had no validity.

He demanded instant action. As soon as the claim was legalized, he walked out to the Shinkolobwe kopje. The pole and the plate still stood, and the mine was now Union Miniére property. What if he had not come back? Suppose he had been killed in action in France? He waited for a quarter of a century for a clear answer to that question.

In 1920 Major Robert Sharp married. He stayed another year in Katanga but found the climate too hard for his wife's health; they went to farm in Rhodesia. Not until 1949 did he return to Shinkolobwe, on a visit of nostalgia. He found the kopje gutted. In its place, behind rusting barbed wire and armed patrols, was a deeply gouged pit with adits giving access to pipes of uranium diving into the ground, a thousand feet deep.

His guess of great wealth had long been verified. After the flow of radium had come the great bonanza of uranium; and associated with this geologists had found an astonishing variety of mineral traces.

Major Sharp had his answer: Had he been killed in France the great wealth might have lain untapped—and there could have been a different ending to World War II.

LONDON—
1920.

The invitation to Rutherford from the Royal Society to deliver the annual Bakerian Lecture came at a crucial stage of nuclear research. In his second year in charge of the Cavendish, his insight had developed new notions of that "strange no-man's-land" he had foreseen in 1903.

In his own laboratory, one of his associates, Dr. F. W. Aston, had invented an instrument—the spectrograph—which could measure the masses of the isotopes and atoms with considerable precision. A similar step had also taken place in America, where the spread of nuclear science in universities was beginning to make headway. In Chicago, a Canadian-born physicist, Arthur Dempster, was measuring atomic masses with increasing skill and calculating the energy binding the particles together within the nucleus.

Considering the text of his Bakerian lecture, Rutherford could estimate the enormous forces within this nucleus to be at least a million times greater than the power that kept the electrons in place. The new measurements, proving Soddy's isotope theory, and adding more definition to the periodic table, gave new leads to the Einstein equation on the equivalence of mass and energy.

By tradition, the Royal Society's Bakerian lecture had been used to review the state of research, looking back over published work. Rutherford could consider the recent understanding of fields of energy in the atomic nucleus which propelled the alpha and beta particles, that the electric charge on the protons in the core was canceled out by the exact opposite charge on the whirling electrons; that this left an atom wholly negative, with the behavior of the electron cloud determining the chemical nature of the atom; but there were still tantalizing, fascinating mysterious components inside his old "no-man's-land." He decided to use the lecture to look into the future; and he engaged in prophecy. In so doing he identified the building block of the universe, and he put his finger on the key to atomic explosions.

He suggested that the basic hydrogen atom—of mass one—was the original constituent of which all atoms, up to uranium, were formed. The nucleus of hydrogen was a proton; the nuclei of all atoms contained protons and electrons, and he suggested that these came together to form other units of matter. This idea made possible ". . . the existence of an atom of mass two, carrying one charge—which is to be regarded as an *isotope of hydrogen.* In the other case, it involves the idea of the possible existence of an atom of mass one which has *zero nuclear charge.* Such an atomic structure seems by no means impossible. . . ." (author's italics.)

The idea of a neutral particle was born.

"Such an atom would have very novel properties. Its external field would be practically zero, except close to the nucleus, and, in consequence, it should be able to *move freely through matter.* Its presence would be difficult to detect. . . . It may be impossible to contain in a sealed vessel. On the other hand, it should enter readily the structure of atoms, and may unite with the nucleus, or be disintegrated by its intense field, resulting possibly in the escape of a charged hydrogen atom, or an electron, or both. . . . It is the intention of this writer to make experiments to test whether any indication of the production of such atoms can be obtained. . . ."

It was the development of his 1904 idea of uncharged particles of high velocity, but far wider based, far more sophisticated.

In the 1920 Bakerian lecture he erected a signpost that would guide the work of the next 20 years. He predicted the existence of the means he had feared in 1916 that would unlock the energy of the nucleus "before men had learned to live in peace with their neighbours," and more. He went beyond unlocking uranium's powerful forces to predict the existence of isotopes of hydrogen (two-part and three-part hydrogen) that would form part of the fuel for superbombs, the thermonuclear weapons—the H-bombs.

It was the most brilliant prediction of his career. He was not to live long enough to see the specter it raised. Unlike some of his pupils, Rutherford was spared the anguish of knowing his talent was prostituted to weaponry.

Within his domain at Cambridge, the hunt was maintained for his elusive, neutral sliver of matter. And his chief researcher was a gifted, reserved English physicist who was not to give up seeking for another 12 years. His name was James Chadwick, then a man of 30, savoring freedom after being interned in Germany for the war's duration, where he had been trapped in 1914 while working with Hans Geiger.

BELGIUM—
1920.

The decision was a matter of arithmetic. Directors of the Union Minière Du Haut Katanga, meeting in the boardroom of the grim stone building in the heart of Brussels, did their sums and reached an easy verdict. The Shinkolobwe mine would be worked; the uranium ores would be shipped to Antwerp; an extraction plant would be constructed as soon as possible. The uranium was a commercial commodity—but only for the radium it contained. There was a fortune waiting on the top of the kopje in the center of Africa. There were rocks there that could be worth two million dollars each.

Radium, by then, had an aura of magic. After the war, the story of the Curies, the blue-green glow, its capacity to fight cancer by searing away tumor tissue with its piercing radiation, lent radium a mystique quite divorced from its mysterious source of energy.

Coveted by competent research groups, it was extremely expensive—when it could be obtained. Possession of radium brought prestige and promotion; no self-respecting major hospital would be without it. Radium became big business. The infinitely small amounts brought incredible prices; commercial interests openly claimed special health benefits for radium-impregnated table waters, beauty preparations, thermal springs—all without thought of danger.

Shinkolobwe gave the Belgian company virtual world monopoly. With the element commanding as much as $3 million an ounce, there was a great fortune in the Katanga ores, and by 1936 more than 10,000 tons had been extracted. Uranium? It was merely the carrier to wealth. Radium was radioactive gold dust. Uranium was the dross.

The opening of the world's richest radium mine was simple. There were no shafts to be sunk, and, at first, no tunnels to be bored, no adits needed. Four white officers of Union Minière engaged 200 black men of the Bayeke tribe, gave simple instructions on what to look for, handed out picks, shovels, hammers, and paid them a pittance for their toil in the African sun.

The workers clambered like black ants among the mushito bushes, scouring the top of the kopje, picking up the float ore, selecting the brightest-colored pieces and stowing them in sacks. When they had hand-picked and sorted the surface material, they shoveled away topsoil and exposed Sharp's reef, biting into its crumbling face with their tools; and, again, the selective hand-picking went on, ton after ton.

The sacks were loaded, the workers chanting tribal songs; the buffalo carts rolled away over the winding dirt track to the railhead at Kambove. In rail wagons built in Britain, the world's first cargo of commercial uranium was carted over the long line to Lobito on the coast of West Africa (Angola) for shipment to Antwerp.

Later, the carts were replaced by hard-tired trucks; and steel drums took the place of the sacks.

Through those first years the laborers from Shinkolobwe sorted and picked the uranium-bearing rocks; then, as the picks and shovels bit deeper and deeper to feed radium to the outside world, geologists found a second vein, ultrarich with showings of cobalt, nickel, native silver, uranite, phosphates, carbonates, molybdates; on and on the list went, until the total of 67 different minerals were identified in Sharp's deposit. And, as Klaproth had first found, the uranium was always associated with lead. It was now showing clear; the family links with uranium went down through generations of elements, through the radium-thorium stages, past polonium and bismuth, to stable lead—a heavy metal like uranium itself, but inert, with the activity, the energetic radiation, exhausted. From element 92 on the periodic table to element 82: here was the evolution of elements—which the Curies had glimpsed a quarter of a century earlier—confirmed!

In Belgium, careful plans were laid to convert to industrial scale the tedious laboratory experiments which had first revealed radium. At the little center of Oolen, near Herenthals, 20 kilometers east of Antwerp, Union Minière's extraction plant was constructed. Madame Marie Curie was appointed consultant to the project, the first commercial production of radium. She had an office in the Union Minière building which she used when able to leave her Radium Institute in Paris. At Oolen, she advised and guided the processes of treatment of the hand-picked ore from the Congo, spending weeks at the site, overseeing the powdering of the ores, and leaching with soda, the sequence of fractionation and crystallization, easing out the traces of radium.

By her own written account, she saw six grams of the precious substance produced—sixty times as much as Pierre had shown to Rutherford in the Langevin's garden. As the processes became routine, the radium factory settled down to extracting between two to three grams each month; and the first eight grams, Marie reported, went as gifts to Belgian centers of radioactive research and for treatment of cancer at various European hospitals.

Such generosity was possible with a price fixed by the Union

Minière tycoons, whose world monopoly lasted until the discovery of extensive fields of uranium (not nearly as rich as Shinkolobwe) at Bear Lake, Canada, in the mid-1930s.

Through the years, at Oolen, the production ratio of radium was as regular as a clock. From each three tons of refined, handpicked ore a single gram of the powerfully radioactive substance was won.

Predestined for some center, it was shielded in lead, guarded and cosseted all the way to its destination. Even residues from containers in which radium, and its cousin polonium, had been held, had a market value to some research centers.

In the field at the back of the radium factory near Antwerp, the waste was dumped. Three tons of uranium concentrate for each gram of radium produced was thrown onto heaps. Yellow, caking in the weather through the passing seasons, it was virtually forgotten. Only a ton or two a year was sold for glassware tinting. The mounds grew to small hills, but drew little attention. Experiments were yet to be made in Paris, Berlin, Copenhagen, to show how these dump heaps were the most strategic material on earth; and those experiments waited for the realization of Ernest Rutherford's prophecy.

Radium, sprung from uranium which had lain for 600 million years in the African heartland, spread across the Western world. Its ceaseless beams of alpha and beta particles, and gamma rays, were prized in a growing number of nuclear research centers, many of them led by Rutherford's protégés. With radium's lance, they probed and tested the buildup of the atom's secret structures.

In Copenhagen, in 1920, the Institute of Theoretical Physics had been established. As its head, its unquestioned leader, was Rutherford's former pupil, Niels Bohr; of him, an eminent associate, Robert J. Oppenheimer, was to remark: "his deeply creative, subtle, and critical spirit guided and transmuted" a synthesis of theory during the third decade of the century.

There were others of note; in Berlin, Otto Hahn had "survived the long interruption" inflicted by World War I. As Bohr opened his work in Copenhagen, Hahn held his first professorial chair at the Kaiser Wilhelm Institute in Dahlem, Berlin. Lise Meitner had taken a lead from Marie Curie and worked during the war with X rays, for the Austrian army. They resumed their pursuit of the radioactive family of elements springing from uranium; but they did not return to enjoy riches from the Oolen radium production. They were forced to turn back to basic stocks.

They dredged residues from previous precipitations of pitchblende, from Joachimsthal, and with this, and with the better facilities and accommodation at the Institute, they followed the habit of earlier days of preparing experiments themselves, without assistants. Like Marie and Pierre Curie, their fingertips were burned and wrinkled, but they soon took more stringent precautions. Slowly, the menace of radiation was being learned.

In the 1920s, Lise Meitner won admiration from her colleagues, and was allowed access to some laboratories where men worked. And she had an admiring family in Vienna, where her sister, a gifted concert pianist, had an ambitious son. He rejected a career in music to which he had been drawn from the time he played Bach, at the age of eight. Now, at Vienna University, he began the work that led to a doctorate in 1926, for a thesis on electrons causing discoloration in salt, a work to be important to radar screens. His name was also Otto—Otto Robert Frisch.

In Paris, at this time, Pierre Bicquard and Jean Frédéric Joliot consulted their former tutor at the School of Physics and Chemistry, Professor Paul Langevin. He warned them: "It is very important you realize that the pay for teachers and researchers is far below what industry offers .̈. . and there is something else! It is a serious handicap in French science not to have been educated at the highest establishments. You will find it difficult to make your way unless you do exceptional work. . . . Think about it for a day or two, then come and see me again."

When they again came to see him, Langevin smiled and said kindly. "I have good news. You, Pierre Bicquard, if you wish, can join my staff as an assistant. And you, Jean Frédéric Joliot, are to see Madame Marie Curie tomorrow morning."

She was small, her hair gray, her eyes still bright; but, the marks of radiation-induced leukemia, the decades of selfless work, were plain, She was fading, even then, in 1925.

It was daunting. He was still in a soldier's uniform, and he was nervous, sitting across the desk from her in the Radium Institute at 11, Rue Pierre Curie, off the Rue Saint-Jacques. But she was as gracious as she was perceptive:

"I like your approach to science. You can join me as a personal assistant on Professor Langevin's recommendation." She looked at his uniform and added, briskly: "Can you start tomorrow?"

He told her he had three weeks' national service to complete. She waved that aside. "I will write to your colonel."

Frédéric Joliot faced more difficulty than had been anticipated. It was an age of snobbery in the starved ranks of scientific aspirants; France treated her young researchers like beggars. Those whose parents could pay their way, who had been given higher education, ignored pushy intruders. Joliot was disregarded, at times ostracized—even after he met Madame Curie's gifted daughter, Irène, fell in love and married her. There were remarks and asides on the "pretty boy wheedling his way above his station in life." Few could find the charity to see genuine affection in the match. He was an opportunist and Irène was disliked as a "block of ice." Her cold manner, her serious face, her resolute application repelled fellowship or camaraderie. Her mother's eminence was another barrier. Frédéric Joliot was to say of her: "I had read much of Pierre Curie; my teachers knew him, talked about him, and I found in his daughter the same purity, good sense, and humility. An extraordinary person, sensitive, and yes, poetic!—a living replica of her great father."

She was a true scholar. She was at ease discussing physics, chemistry, poetry, in French, German, and English. She worked hard, determined to emulate her mother. Joliot was persuaded to take the name Joliot-Curie, so that her father's name should pass to their children. Together, they began a career at the Radium Institute—later to be resumed at the Collège de France.

When the Joliot-Curies opened their joint research in the late 1920s, radium ruled above other means of firing alpha projectiles at the atom's core. Soon, though, it lost its crown.

Advances in understanding the origin of radiation, of its wave-like patterns, of the strange fields of force which held the nucleus in extreme density and kept the electrons apart, made it clear that to break into those secret halls would call for armament of higher energy than Rutherford's alpha shells. Their rate of fire was too low, their penetration too weak, to smash into the nucleus itself. They were always flung aside. If the construction of the nucleus was to be revealed, there would have to be new artillery, more highly powered projectiles to strike through the atom's bastions.

Many minds struggled with the problem. George Gamow came to the Cavendish from Russia with an idea; and two of Rutherford's workers, John Cockcroft and Edward Walton, spent 500 pounds (a huge sum in those days) on materials to build a device to accelerate a stream of protons impelled by a power surge of 500,000 volts. This outnumbered radium's alpha particles. Bombardment by energy-laden protons was prodigious; it equalled alpha radiation from thousands of grams of radium. With this new-style atomic cannon, the two men hit the cores of lithium atoms and broke them into two parts—which turned out to be no more than two alpha particles! It was deliberate, man-made transmutation. Cockcroft rejoiced, "We've done it! We've split the atom!"

Splitting the atom was soon a catch phrase, redolent of the ultimate achievement.

They had split an atomic core, as Rutherford had done in 1919; now they could measure more precisely the energy of liberated fragments. They could assess the strength of those mysterious, binding energies, and their result gave accurate verification of Einstein's law on the equivalence of mass and energy. The door to the nucleus of the atom had moved ajar.

The view of the central fount of all energy in the universe was a mere glimpse, however. The mystery of the components remained. Klaproth, Roentgen, Becquerel, Mendeleev, Rutherford, the Curies, Hahn, Geiger, Bohr—all the pioneers—had traveled a path of astounding marvels. The concept of solid atoms had been obliterated.

Now, on the threshold of the 1930s, they were at a watershed where the appearance of Rutherford's fragment of neutral matter offered a wonderland of new exploration—and opened a path to atomic war.

SECOND PHASE
The Invisible Particle

One Tuesday in February, 1932, about twenty of the Rutherford team are crowded into Fowler's room at Trinity, all of them members of an exclusive group known as the Kapitza Club. They meet, usually, twice a month during term, always on Tuesday evenings.

It is not a club, as such—no officers, no rules, although John Cockcroft, always precise and neat, likes to keep minutes of the gatherings. Kapitsa started the meetings in an effort to break down the barriers of official research, to allow an exchange of ideas which few people outside their circle would understand.

Pyotr (now Peter) Kapitsa was the son of a Russian general, gregarious, eccentric, with a bold, questing spirit Rutherford admires greatly, and a flair with the generation of high-energy currents needed to create powerful magnetic fields to influence the paths of the Old Man's alpha "projectiles."

They sit and talk and watch the door. Outside the cold wind, cutting through Cambridge, rattles the trees beyond Neville's Court, and Professor R. H. Fowler, Rutherford's son-in-law, draws the curtain to shut out the night. They have all wandered in after dinner; Cockcroft is at the small table, and in the groups scattered through the room are other senior men; Blackett is there, tall, formal, chatting with Norman Feather, and Geoffrey Taylor—who will be one of the first men to calculate the energy release from a fission bomb—and J. K. Radcliffe, along with C. P. Snow. Also there is Mark Oliphant from Australia.

All of them know this evening is a special event. Kapitsa is entertaining a special guest at dinner in Hall, at Trinity. Is it a celebration? This is part of the rumor which has flooded the Cavendish. They anticipate an advance in understanding the enigmatic atomic core. They do not yet know that from this evening forward the uranium nucleus will become vulnerable to the attacks of men.

Many of these researchers were still at school when the Rutherford prophecy was uttered. All of them are aware of its implication; the topic has not been bandied about publicly since the Bakerian Lecture of 1920, but for 12 years, James Chadwick has been persistent in hunting an invisible particle which Rutherford said should be there—which has never been sighted.

All through the 1920s Chadwick has lived with the notion as steps in knowledge built up a picture of the nucleus in the imagination. With logic, with more precise means of assessing the masses of atoms—knowing the energies needed for alpha particles to escape—they all recognized there were grounds for Rutherford's theory that "something else" was packed into the center of the atom other than the known protons and alpha particles. But was there really a ghost particle, a neutral fragment that did not reveal itself in the electrical detection nets?

Rutherford, now knighted, has had five years as president of the Royal Society. He still believes in this undetected, elusive sliver of matter, and has confessed this view did not rest so much on the empirical evidence: "It comes to me from long and deep consideration of how it is possible to build up a complex nucleus from the known elementary bits and pieces. . . ."

Kapitsa enters the room, beaming, eyes ablaze with triumph. With him is James Chadwick, quiet, reserved; his face is drawn, and gray shadows under his eyes tell of nights without sleep. Yet he is smiling and patient with them as they press around him, prompting him to speak. And he says briefly, "We've found it. We've pinned it down. . . ." As though it were a butterfly! "It is an uncharged particle, a part of the mass of the nuclei; and, it can move freely through matter, as he said it would, because it is neutral. It has no charge. I propose it shall be known as the neutron."

Rutherford, said Chadwick, had never abandoned the idea of the neutral particle: "He completely converted me to the notion, and, from time to time over the years, we devised experiments to find how it might be formed and how it could escape from the enormous binding forces of the nucleus."

Hans Geiger and a colleague named Müller had devised far more sensitive particle counters in 1928 and these had been used in the search, but without success. Chadwick returned to the scintillas of light, where radioactive particles struck target atoms, just as Rutherford had done with Soddy in Montreal in 1902; they called it the scintillation method. It revealed nothing new. Then, subsequently, a new, powerful radiant source became available; this was obtained from radium tubes that came from the Kelly Hospital in Baltimore. Chadwick set Dr. H. C. Webster, a young Australian researcher, to extract polonium created in this material.

Meanwhile in Germany two workers, H. Bothe and W. Becker, reported bombarding beryllium with a strong source of alpha radiation and wrote that this gave off what they took to be "very hard gamma" radiation, more penetrating than anything known since it would pass through 12 centimeters of lead. There had been two repercussions to this report—one in Paris, the other in Cambridge.

Chadwick had Webster make a similar test, and he found the same highly penetrating radiation; also, he noted, what was emitted in the forward direction was more powerful than the radiation scattered backward.

"I was excited by this fact," Chadwick related. "It could only be readily explained, it seemed to me, if this radiation consisted not of wavelengths, of gamma rays, but of *neutral* particles."

Yet no traces, no tracks could be found to validate the supposition.

In Paris, Joliot-Curie and his wife Irène had also read the paper by Bothe and Becker. Through Madame Curie they had access to the greatest single collection of radioactive sources, including the wealth of a gram and a half of radium. As Chadwick and Webster had done, they extracted the polonium that had grown from the decaying radium over the years. Their source was a far more powerful beam of alpha bullets to fire into the atoms of their target materials of beryllium and aluminum. They screened off normal radioactive fragments with a sheet of lead, and those particles not absorbed passed through an ionization chamber (two plates with a voltage between them, in a gas-filled chamber. Any charged particles passing through would ionize some of the gas atoms and cause a small voltage pulse at one of the plates.) They also allowed the penetrating particles to traverse the pressured air of a cloud chamber and so reveal their tracks by leaving a trail of condensation droplets, which could be photographed. (These particle detectors, invented by a Scots physicist, C.T. Wilson, in 1913, were known as Wilson cloud chambers.) Joliot-Curie, a clever engineer-physicist, had recently constructed several cloud chambers. Thus the couple in Paris were fully armed for discovery, except for one critical essential— they had never read Rutherford's Bakerian lecture; they had dismissed it as a view "of old work reported previously." It was a costly omission for Irène, who, wishing to emulate her mother, had an eye on her first Nobel Prize. The irony of this mistake, for Irène, was that it would be repeated, in another way, and she would narrowly miss the chance not just to match her mother's two Nobel awards, but to win *three* Nobel Prizes, each for historical discovery.

They wrote their paper for the French Academy journal— *Comptes Rendus*—and declared that alpha particles, fired into targets of beryllium, dislodged the atomic nuclei of hydrogen, helium, or nitrogen. Their work was elegant, but their interpretation was tentative. They, too, thought the ensuing radiation might be gamma rays. The Joliot-Curies called the results "new and interesting"; Chadwick told his audience in Trinity: "They reported a surprising radiation from beryllium—that of ejecting protons from matter containing hydrogen ... a most startling property! Soon after, Feather came to my room to tell me about this report—as astonished as I was!"

It was Chadwick's custom to visit Rutherford's office each morning to talk over work in progress in the laboratories; he told him of the French report: "I saw his growing amazement. Finally, he burst out: 'I don't believe it!' Such an impatient remark was utterly out of character. In all my long association with him I can recollect no similiar outburst; I mention it only to underline the electrifying effect of the Joliot-Curie report. Rutherford, of course, had to believe the observations reported; the *explanation* was quite another matter. . . ."

Chadwick now had a "beautiful" source of polonium, plucked from the radioactive tubes from the Baltimore hospital, and, with this, planned his own assault on the mysterious radiation.

"I started with an open mind, though, naturally, my thoughts were on that projected neutral particle. . . . I was convinced there was something new, as well as strange. A few days (and nights) were sufficient to show these strange effects were due to a neutral particle—and to enable me to measure its mass. The neutron, which Rutherford postulated in 1920, has—at last—revealed itself."

How could it have been found, be shown to exist?

Rutherford had resorted to H.G. Wells.

"How could you find the Invisible Man in Picadilly Circus? You couldn't see him! But, you'd know he was there by the people he collided with, by the reactions of those he pushed aside. Recoil, my boy! That's how the neutral particle gives itself away . . ."

It was Chadwick's triumph; it was Rutherford's final contribution to the conquest of atomic mystery. Without his foresight, without the neutron, there would have been no atomic project in America.

The neutron was the final key to wrenching nuclear energy from uranium. Rutherford was never to know that; for him, it was just as well: when he died in 1937, fission was still two years away.

He had shown atoms were not solid, that "spontaneous disintegration" happened inside uranium and its radioactive family. He showed the atom to be planetary and its space filled with strange forces. He was the first person to achieve transmutation, and he prophesied the existence of the neutron and two critical isotopes of hydrogen—to be called deuterium and tritium.

A co-worker of Rutherford christened that year of 1932 *annus mirabilis,* and a year of wonders it certainly was. As Cockcroft and Walton "split" their atoms with the Cavendish accelerator, so a brilliant 31-year-old American, Ernest Orlando Lawrence, working in California, was building the world's first cyclotron—a machine that would give an energy of 1.2 million electron volts to a stream of protons. In that same year, Harold Urey and two American associates found deuterium, the projected two-part hydrogen—a double-weight isotope that forms with oxygen into molecules of water to make "heavy water," the substance due to play a role in the future generation of atomic energy. ("Normal" hydrogen, of course, also unites with oxygen—to make "ordinary" water.) Only two years later, Mark Oliphant, at the Cavendish, discovered Rutherford's predicted "three-part" hydrogen, which he called tritium. It, too, was destined as fuel for a terrifying weapon.

• 2 •

Since Rutherford's discovery of the atomic nucleus, the small community of atomic physicists had been trying to create a picture of a miniature universe; now the structure of the atom was approaching comprehensibility. It seemed fitting that in 1933 this should be announced in Brussels, the seat of power over the Shinkolobwe uranium mine, from which the major supply of experimental radioactive substances flowed. Yet it was not held here because of Union Minière interest, but rather because Ernest Solvay, a Belgian chemical tycoon, had financed this seventh Solvay Congress.

More than 40 of the top workers in the nuclear field came, from Russia, Britain, America, Italy, Denmark, France, Austria, Germany. Among them were Madame Marie Curie, and, beside her, the chairman of the proceedings, Paul Langevin, the man with whom the gossips said she was having an affair. At her other side was the Russian, A. Joffe, and next to him sat Niels Bohr. Beside Bohr was Irène Joliot-Curie. Behind her stood her husband, Frédéric Joliot-Curie.

There also were the twin victors of the split atom, Cockcroft and Walton, and the inventor of the cyclotron, Lawrence, from California. Another Rutherford collaborator, P. M. S. Blackett, the former naval officer who was to capture more than 400,000 separate photographs of particle trails in cloud chambers, was standing next to the Russian-born George Gamow, and nearby was the Italian Enrico Fermi, soon to be the first to shatter the uranium core and not know it.

To the left of Langevin sat Ernest Rutherford and, near him, Bothe of Germany, whose work had led to the finding of the neutron; then there was Werner Heisenberg, who was to lead German nuclear work in wartime. Also in the group was James Chadwick. Opposite to Irène Joliot-Curie was the close associate of the nuclear chemist, Otto Hahn, of Berlin. She was the Austrian Jew, Lise Meitner.

For the first time at such meetings, events in the outside world were having an impact on atomic science. Lise Meitner's nephew, Otto Frisch, had been forced to leave his research post in Hamburg to work in England; he was never again to work in Germany. That year Hitler had come to power amid bloody riots; a Dutch Communist, Marinus van der Lubbe, was made scapegoat for the burning of the Reichstag building, and decapitated; Jewish researchers were being hunted from their posts, and Nazi Germany quit the League of Nations. Fear of racial repression loomed over the discussions.

And there was another influence new to all scientific symposia: the neutron. It guided their talk, their questions; and its discovery, of itself, brought acrimony to the exchanges.

Weeks before, Joliot-Curie had written to a confidant in Russia, D. Skolbeltsyn, displaying chagrin and some blindness to his own animus: ". . . it is annoying to be overtaken by other laboratories which immediately take up one's experiments. . . . In Cambridge, Chadwick did not wait long. . . ."

Joliot-Curie had told his Russian correspondent, "We have been working hard—we had to speed up the pace of our experiments. . . ."

He and Irène had fought disappointment and had taken the work of bombarding beryllium further in the Radium Institute in Paris. They had a report on this (to be read to the Solvay talks) which was headed "Penetrating Radiation From Atoms Bombarded by Alpha Rays." They claimed to have found, with emerging neutrons, the newly reported positive electrons. Nobody disputed the existence of positive electrons—positrons as they came to be called—since they had been detected in showers of cosmic radiation in the earth's atmosphere. But neutrons and positrons from light elements? Was this interpretation valid?

Lise Meitner took the floor. In the Kaiser Wilhelm Institute, she coldly declared, she and Otto Hahn had performed this very same experiment.

"We have never found anything but the emission of a proton."

The weight of discussion swung against the two workers from Paris. It was a shattering experience for Irène; it founded the acrimony that grew between the center in Paris and the Hahn laboratory in Dahlem, Berlin. The relationship between the two women in the next years was the spur to a startling development.

· 3 ·

The neutron was not an instant magic wand. But, its effect on physics was vital as a new and powerful tool of bombardment, as a fresh indicator of what was inside the incredibly small and difficult nucleus. It was not yet seen as the trigger for the liberation of atomic energy. It was to be five years before the historic importance of the neutron could be comprehended.

The Joliot-Curie results were part of confusion stemming from incomplete understanding of the atomic core. The disappointed couple left for Paris. That was science, said Langevin; when anything new came, people did not want to believe what they had not found themselves. The kindly Niels Bohr told them he thought the work important, and the Austrian, Wolfgang Pauli, had consoled them. Soon they were to receive a letter from Rutherford, but that came *after* their rehabilitation.

They worked together each day in the winter months, breaking only for Christmas with their children, Hélène, and the baby son, Pierre. They sought answers to their mystifying results. Was this akin to Rutherford's McGill experience when firing alpha projectiles? Was it some kind of instantaneous disintegration or disruptive change in the target atoms? Was there some unknown mechanism at work? Was the speed of their projectiles greater than Hahn and Meitner had used in Berlin? Did that make the difference? They were certain they had not made a gross error; still, they repeated their work slowly, carefully, guarding against assumption.

It was mid-January when Pierre Biquard was telephoned by Joliot-Curie: "Please come over to the Institute, at once! There is something you must see."

The equipment covered two tables in the basement laboratory. There were hand-built cloud chambers; there were screens of aluminum and beryllium, detectors and mechanical counters, ionization tanks, wires, and coils—and, standing by, a smiling Joliot-Curie and a serious-faced Irène. The door opened and Marie Curie entered on the arm of Paul Langevin; she smiled slightly, and waited, silent, as Joliot-Curie prepared the demonstration and reviewed the tests made following the Bothe and Becker work.

"We have gone back over that whole ground; we came on an explanation to-day," he said. "We have discovered a new atomic phenomenon. Let me show you."

He exposed the polonium source so that its beam of radiation struck the beryllium. In the silence the Geiger counter crackled into life.

"That is as we all expect," he said. "And, when I close off the radiation beam, then we would also expect the counter to stop recording the passage of particles. But, it doesn't! Listen. . . ." The click-click-click continued.

At first, they had questioned whether their apparatus was faulty, and they asked a young German student, Wolfgang Gentner, to make an independent check of the counting equipment. He found it in perfect order.

Joliot-Curie announced, proudly, "We were not mistaken on the detection of positive electrons and neutrons from the transmutations we attained. More importantly, we now know that positrons continue to stream from the target materials—from aluminum, from beryllium—and from magnesium! It means—and we have checked this chemically—we have made the first *artifical radioactive materials*. There is a broad prospect ahead. We should be able to produce these isotopes, and others, by irradiating matter with various particles."

He did not yet know exactly what had happened, only that the first radioactive isotopes had been made by man. Here was the last great thrill for Marie—the triumph of a daughter and son-in-law that would earn them the Nobel Prize for chemistry in 1935. Sadly, Marie would not survive to see that event. She died on July 4, in Haute Savoie. Frédéric Joliot-Curie remembered how "Irène and I showed her the first artificial radioactive element in a little glass tube. I can still see her, taking in her scarred fingers this little tube containing the compound. Though it was weak, she held it close to the Geiger counter and she could hear the rate-meter giving off a great many clicks. It was the last great satisfaction of her life, and I could never forget her intense joy."

The Joliot-Curies could only glimpse the outcome of their findings. Their first three radioisotopes were to grow to a thousand different types, to be born in neutron storms in the coming age of reactors. Eventually, isotopes prepared from dozens of elements and compounds allowed use of tracers as diagnostic substances that telegraphed their positions in the human anatomy, in plants, in wild life; they gave direct application in the treatment of illness, in pathology, to biology, and biochemistry. An investigative art developed that broadened the whole stream of science and made possible immense advances in human knowledge.

· 4 ·

Twenty years before the neutron was discovered, the fertile brain of H. G. Wells saw a vision of an atomic world. In 1913 Wells' book was published throughout Europe, in several languages. Among his lesser-known works to-day, *The World Set Free* was one of his most accurately prophetic works. With only the facts of natural radioactivity (that uranium changed the emission of particles) and Einstein's theory of equivalence of energy and matter to guide him, Wells pictured future cities run entirely on atomic energy. He had trains, aircraft, ships driven by this mysterious new power—and, of course, told of a war with atomic bombs that brought it all to an end.

With uncanny accuracy he set his tale two decades ahead—to 1933—the year of the neutron and of the first glimmerings of the atomic chain reaction.

Wells' vision of an atomic world fired a few young minds. In Budapest, Leo Szilard, at the age of fifteen, committed himself to a career in physics. Like Wells, he was to be a prophet of science. At the age of twenty-four he gained his doctorate—under Planck and working with Lise Meitner. He was in the faculty of Berlin University for eleven years, until he and many others, including his townsman, Edward Teller, knew there was no safety in Germany. Teller sought sanctuary in Copenhagen, and Szilard went to England.

In Szilard's private papers, found after his death in California in 1964 . . . he had described how, when in London in the autumn of 1933 Rutherford had been reported as saying it was "moonshine" to hope to get industrial power from the atom; that had been a red-rag challenge to Szilard. He sought ways to prove Rutherford wrong. He was walking down Southhampton Row toward The Aldwych, and, where trucks came out of Covent Garden with loads of fruit, vegetables, and flowers, one of the first sets of traffic lights had been fitted. Szilard waited for the light to change from red to green. He did not move. The thought was mesmeric: green to amber, amber to red—one light, two lights, three lights. His thoughts raced, circling around the idea of the behavior of the atomic nucleus. And the theory was born. After all, there was the high-energy, penetrating neutron—not the best atomic artillery known—and you fire it direct into the heart of another atom. And what if you found an element in which the nuclei threw off energy, as Rutherford had shown some elements did quite naturally? What if you could *make* it happen, at will? What if this element's atoms threw off two new neutrons to strike two more? Two twos are four; four fours are sixteen; sixteen sixteens are—in a flash the figure would be astronomical. With Einstein's law, matter released as energy, there would be in such a continuous process ample power liberated for industrial use from the atom. Moonshine? All you need to do is to find the right element! And who to talk with about that?

It was Szilard's style to go right to the top. He was to show this, in 1939, in his dealings between Einstein and President Roosevelt; in 1933, the top man in physics was in Cambridge. Szilard took a train, went to the Cavendish, and confronted Rutherford in his office. Now in his sixties, white-haired, Rutherford came out of the 20-minute meeting with the Hungarian almost speechless; wagging his head and muttering, he stalked the corridor, aching to burst out his feelings to somebody, anybody. Along the passageway came a young American student who counted it his "good fortune to have been admitted to the Cavendish" that summer, by Rutherford, who now fixed him with his angry eye and told him of Szilard's "harebrained idea" of shooting neutrons into light elements, like beryllium, to get surplus energy from a chain reaction! Absolutely impossible, totally impracticable; an outlandish proposal!

It was the first time in his life the young American heard the expression—chain reaction. He was never to forget. He was in the desert of New Mexico in 1945, responsible for firing the first atomic explosion, and later, he became professor of physics at Harvard. His name was Kenneth T. Bainbridge.

And Szilard did not forget. Not a man to relinquish ideas quickly, he worried at the concept all through the following winter and spring. On June 28, 1934, he filed an application in the London Patents Office for the world's first registration of a nuclear process—chain reaction by neutron bombardment—a move which gave Rutherford further offense, since he was firmly opposed to commercial gains from research. However, Szilard had reasons other than personal profit. That year Hitler proclaimed himself president and chancellor, and assumed the new title of Führer.

Szilard was afraid his chain reaction idea could fall into wrong hands. To keep it secret he assigned British Patent No. 630726 to the British Admiralty. It was the first of two occasions when naval authorities had the chance to develop the most destructive weapon. Szilard's patent rested in a dusty pigeonhole in Whitehall. It had one serious flaw; it proposed neutron bombardment of the light, brittle metal, beryllium. He chose an element at the wrong end of the periodic table,

· 5 ·

London seemed a lonely place to Otto Frisch. A slight, quiet, shy man, worried as well as solitary, his manner did not encourage warmth and sympathy. He passed his days in Blackett's laboratory and, although Blackett was polite, Frisch could not forget he was an exile. The news from home was ominous. Through 1933 concern had mounted in letters from his parents in Vienna and from Berlin where his aunt, Lise Meitner, still worked on the search for radioactive substances. Rumor said the repressive laws against Jews, which had hunted him from his university post in Hamburg, were being more widely applied. True his aunt was an Austrian national, but she was still a Jew; and Austrian nationality had not saved his position in Hamburg.

In 1934 his concern deepened. The winter was hard, and in his little room he had no piano with which to find solace in Brahms, Beethoven, Liszt.

Britain was still struggling with the Depression and wide-scale unemployment, and, in the misery, London was drab, with only the exciting news from Paris and Cambridge lifting his spirits. As did many young physicists of the day, Frisch rushed at once into a string of bombardment experiments, using alpha particles—all directed against lighter elements, but with no startling or unusual results. There seemed nothing to brighten his future until the offer arrived.

The Danish stamp on the letter brought back a warm memory of the previous year, of writing to his anxious mother from his room in Hamburg: "Please don't worry about me or my future, mother dear. I know things look black for our people all through Germany, but. . . ."

He had met Niels Bohr; the Dane had been prowling through

German centers, assessing the savagery of the Nazi regime on Jews and researchers who refused to swear allegiance to the Nazi Party, trying to gauge how external help could be organized, and he had come when young Frisch had just measured accurately, for the first time, the incredibly tiny force of the recoil of an atom which had ejected its quantum of light.

"It was a wonderful experience, mother, to be confronted, suddenly, by this big, almost legendary figure, and to see him smile at me, like a kindly father. And he caught me by one of my waistcoat buttons, drew me nearer, and said—'I hope you can come and work with us sometime. We like people who can think up such experiments.' That is what he said to me. . . ."

Now Bohr's offer was in his hands: a position at the Institute of Theoretical Physics in Copenhagen; a place to work in the laboratories and a room to sleep in under the roof of the building at Blegdamsvej 15; a marvellous, reinvigorating chance to work among some of the finest brains Bohr had culled from Germany; to work in a relaxed atmosphere, after the stiff formality of the Blackett laboratory! And he could build precise, more simplified particle counters and auxiliary equipment to hunt for secrets among the radioactive substances.

From London he wrote to his mother, "Do not be concerned any longer—the Good Lord has taken me by the waistcoat button, and smiled at me."

· 6 ·

The first deliberate attack on uranium with neutron projectiles took place in Rome. It happened there because a dynamic physicist named Enrico Fermi needed a change of study. Professor of physics at the University of Rome, he was enticed from his work on rays by Chadwick's discovery of the neutron. A neutral particle that could pass through matter excited Fermi's sharp imagination. Then, when the Paris work produced artificial radioactivity, it was clear to him that here was a method of pushing extra fragments into the heart of the atom, a chance to change the course of his work to a less crowded, more promising avenue.

Fermi was among the more unusual of nuclear explorers.

Born in Rome in the year Rutherford was discovering that atoms
of the uranium family were spontaneously disintegrating, he won
his doctorate at Pisa at the age of 21 and went to study in
Germany. Married to Laura, daughter of a well-known Jewish
family in Italy, he loved his two children and sports; indeed, it
was while playing tennis that he made the decision to use the
nuetron as a powerful projectile to assault atomic structures. He
was to be the first man to achieve fission, but he was not to know
that until it happened elsewhere.

Fermi had a team of keen young workers. He had such a gift
for seeing into the heart of problems and a paternal manner of
solving them—with a piece of chalk in his hand—that his collaborators irreverently named him The Pope. His chief colleague,
Emilio Segré, became a lifetime friend, and, like Fermi, was to
play a role in the development of atomic weaponry; and the
brilliant Oscar d'Agostino, a young physical chemist who had
been with Madame Curie's Radium Institute in Paris, came
home on holiday and stayed on. Five of them—with Amaldi and
Rasetti—were all to have major impact on scientific history.

Fermi made his decision on the new line of work with this
comment: "I am opposed to the seeking of new radioactive phenomena on the basis of random guesswork. The only chance now
of great discovery in atomic physics, after all that has been done,
is in the *modifying* of the nucleus of various atoms. That will be a
worthwhile goal for the future."

It was an error of approach that cost him the most significant
discovery of modern times; but almost as certainly it saved the
world from the terror of a rampaging Nazi Germany armed with
atomic bombs.

The Fermi approach was essentially systematic. He began at
the bottom of the periodic table, planning to work his way up the
scale of increasing mass of the atoms of the various elements. He
got no results until he bombarded the ninth element, fluorine;
this became artifically radioactive, just as magnesium, lithium,
and boron had in Paris. Steadily, Fermi worked on, adding more
artificial isotopes to the store that would be used to explore the
physical world and the processes of living matter.

When the first news of his work was out, many physicists—
including Frisch—rated the Fermi concept as silly. Imagine the

bother of producing neutrons for this work when alpha particles were far more numerous! This attitude exposed the limited knowledge of atomic structure of the day; only the neutron could spear into the nucleus and be impervious to the fierce fields of electrical repulsion defending the core of the atom.

Fermi worked his way up the periodic table until he faced the heaviest of all natural elements—uranium. All but 18 of 60 different substances bombarded had been made artificially radioactive. What would the result be of such an assault on nature's most compact nuclear citadel?

It was reasoned that, since atoms by absorbing neutrons could move up the periodic table to become atoms of a slightly heavier element, to detect any change in the uranium nucleus to something heavier—by adding a neutron—would need a screening out of the normal radiation. Anything new created by the jump over nature's fence into the ground beyond uranium—Fermi coined the word *transuranic* for this state—would mean a fierce kind of radiation not detected hitherto.

"We shall erect a screen," he told d'Agostino, "that will serve to blot out the short-range particles and allow radiation from the new element to pass through."

The screen was a sheet of aluminum foil, only three thousandths of an inch thick. It was enough to conceal the truth. The foil hid the fragments of the explosion of certain atoms of uranium struck by a neutron. Fermi and his team looked in the wrong direction. They looked *up* the periodic table; then stared *over* the "transuranic" fence. And they stood in puzzled amazement at the confusion of chemical substances that d'Agostino found in his crucible after treating the irradiated uranium.

Instead of a single new element in the dish, he detected 20 active substances, all reducing their radioactivity at different rates, all with varying half-lives—and of a nature and origin not yet explained. Fermi repeated the process; however, this time he did something different. As before, he aligned his glass tube of radium so that its alpha particles struck a sample of beryllium— in effect making it a neutron gun—but between this and the filtering foil screen he now placed paraffin wax. Any material containing hydrogen would have had the same result. The fast-flying neutrons were slowed down and, being slower, struck a larger number of target nuclei—a fundamental discovery that

Fermi was to use in building the world's first self-sustaining nuclear reactor.

Still the sights in Rome were unchanged. Fermi was seeking heavier transuranic elements; beyond element 92—uranium—would be a 93, or a 94, perhaps a 95! He continued to seek a new creation that, at best, would have been a few atoms among trillions; and so his thinking was guided into assumption. Among the unusually radioactive debris d'Agostino found in the target material, Fermi asserted there *were* transuranic elements. He was convinced. He published his results with that conclusion.

In Berlin, Otto Hahn and Lise Meitner eagerly followed Fermi down the transuranic path. Their work showed Fermi's observed radioactive substances would not be linked with any atoms immediately *below* uranium—right down to mercury, element 80. Hahn was solid in the view there was no alternative to Fermi's claim to have found transuranic matter. Lise Meitner, however, was uneasy about the conclusion. She did not then have the sagacity of another German lady, Ida Noddack, already an acknowledged expert on the elements known as "rare earths." She was of some standing; for, in her earlier work, she had discovered the new element rhenium, element 75, which fitted into the Mendeleev table, between wolfram and the exceedingly dense metal osmium.

She wrote to Fermi and sent him a copy of a paper, published in a German chemistry magazine; in which she said, "It is . . . possible that, under neutron bombardment, heavy nuclei, like uranium, break into several large fragments which are actually isotopes of known elements, but not neighbors of the irradiated elements . . . ." No claim to transuranic elements could be valid, she said, until *all* the elements had been eliminated.

Shrugging off this brilliant nuclear insight, Enrico Fermi compounded his great mistake and went on seeking transuranic matter. How could it be that a nucleus could be shattered by a neutral particle—which did not have the kick of a single electron volt—when alpha particles boosted to million-volt power were repelled? He had an excuse; the foil screen had shut out the truth of the bursting atoms of uranium. Leading workers in Europe and America followed him up the road to look beyond the uranium fence.

In hindsight, what took place in the nucleus of element 92,

when it was irradiated in Rome and in Paris, Berlin, and the United States, was to seem so apparent that it became tempting to seek some influence acting against a dawn of awareness. Some observers have seen this work—performed in the virtual shadow of the Vatican—as masking facts that would have led to Hitler having the chance to develop atomic weapons prior to World War II.

Was there some influence at work? Did some force guide thinking so that the Western world was given a period of grace—which the American Pulitzer Prize-winning science writer, William L. Laurence, called the "Five Year Miracle"? No doubt, had the fission secret of the uranium nucleus been broached in 1934, German scientists would have recognised the potential in liberation of uranium's vast energy. Would they have joined in a national drive to be first with the atom bomb? Would those talented men—due to play such a role in the American bomb project and not then exiled by tyrannical insanity—have resisted the challenge of harnessing the atom for war?

In the year of the Fermi uranium experiments, Lord Rutherford lost the services of a favorite "son."

Like other Russian-born scientists who were to play a role in the atomic arms race, Peter Kapitsa had left his homeland as a young man, at the time of the Bolshevik takeover. His talent lay in building equipment to increase the velocity of the particle "shells." At that time, the Cavendish was unknowingly fighting against a tide that was to leave it a backwater of the critical uranium study. The same deception that misdirected Fermi's mind and misled the Western world's centers also hid the truth from Rutherford's group. Persisting in using high-speed machines for hurling their shells at lighter target atoms, they all failed to grasp the true nature of the tool that had been discovered in their midst; the neutron was used, but not against uranium!

Kapitsa had a high reputation in science, and in 1934 the Soviet Academy of Sciences elected him a member. He returned to Moscow. Once there, he was informed he would not be allowed to return to England. The argument, it was reported, was that Stalin felt Russia had need of his services because of the fear of the rising belligerence of Hitler's Germany.

Rutherford was furious. He raised every protest possible through diplomatic channels up to Prime Minister Baldwin; it was vital for the progress of atomic science that Kapitsa continue his work at the Cavendish, he argued. The Kremlin replied it was understandable that England would like to have Kapitsa—just as Russia would like to have Rutherford.

In the face of this defeat, Rutherford showed a complete lack of wisdom of how the atom might be used in world politics. He shipped the entire equipment in Kapitsa's Cambridge laboratory to Russia. The Russians paid the Cavendish £30,000 in compensation.

Working under Stalin's wing, Kapitsa became a key brain in the Soviet school of nuclear physics. In the end, he was as committed to advancing Russian status as the exiles in America were to be in developing atomic weaponry, so much so that, later, he made overtures to Niels Bohr to work with him in Moscow.

Against those facts, and the known character of the Hitler and Stalin regimes, failure to accept the truth of Fermi's Rome results in 1934 could, indeed, have been a five-year-miracle for the free world.

· 7 ·

Otto Frisch at the Institute of Theoretical Physics in Copenhagen had a copy of small science journal—*La Ricerca Scientifica*—the magazine with the right to print Enrico Fermi's papers.

Frisch read, and translated, the exhilarating results from Rome. Now, losing his shyness with each sentence, Frisch was the center of attention as he related what had happened in the Fermi experiments.

New radioactive substances had been produced from neutron assault on uranium! Some of these were described to be "beyond uranium"—transuranic elements, man-made! And, not the least thrilling, the neutron—so powerful a tool for probing the mysterious matrix at the center of the atom—was made more effective by first passing it through material containing hydrogen, such as water or paraffin wax.

As each new issue of *La Ricerca Scientifica* arrived, a crowd gathered about Frisch, soaking up details of Fermi's reports. It was soon apparent the Institute needed a strong source of neutrons; this was urged by Georg von Hevésy, Hungarian-born discoverer of element 72—hafnium—so-called from the Latin name for Copenhagen. (Hevésy had worked with Rutherford before joining Bohr and originated the "tracer" method of biological detection, the labeling of known organic matter, or bodily chemicals, with radioactivity so that the progress, and process, in living organisms could be followed. He was the first user of isotopes in medical science, for which he was awarded a Nobel Prize in 1943.)

Hevésy suggested the Danish people should be invited to show appreciation of the great Niels Bohr by subscribing to a present for his fiftieth birthday—and the present was to be a half-gram of radium at a cost of 100,000 kroner! The radium was extracted from Shinkolobwe uranium ores, still being processed at Oolen.

Bohr's radium, however, was not prized for its cost, but for what it would provide—with skilled manipulation. By itself, for all its radiant power, it was not a source of neutrons, and it was neutrons the Institute now needed badly. Otto Frisch was handed the task of transforming the radium to a source of powerful neutron emission. Meanwhile, the radium was stored in a well, already driven into the floor of the basement; and there it rested, little more than one-sixth of an ounce, waiting to be converted to a neutron gun by being mixed into a compound with beryllium.

The Institute was now turned into a modern version of an ancient alchemist's workshop. Frisch needed several grams of powdered beryllium, and for days he and his colleagues worked tirelessly, raising blisters on their hands, pounding the hard, silvery metal in mortars, slowly breaking it down into finer and finer powder. They worked not knowing some people are highly sensitive to the powder, that death could have come from inhaling a speck of it. The beryllium powder was finally mixed with the radium, and, at once, the neutrons were flowing in a copious stream, an unending and virtually undiminishing source.

Initially, the Bohr neutron source was intended also to aid the biological studies at the Institute; Hevésy wanted neutrons for some of his isotope tracer work. It was this which brought the first danger to the center. The radium-beryllium compound was kept in a large glass container, in two gallons of carbon di-sulphide. This was stored at the bottom of the well, and, under the hail of neutrons, radioactive phosphorus for isotope experiments was created in the highly toxic, dangerously flammable solution. It was carried up a spiral staircase from the cellar to the laboratories and, one day, a laboratory assistant slipped; and the bottle fell and smashed. He raced up the stairs with the deadly fumes following, shouting a warning, everyone knowing a single spark could engulf the whole building in a sheet of flame. The scientists rigged the largest pump they could acquire; all the air in the building had to be exhausted for safety, and the pump continued working through the night.

Otto Frisch slept in a small room under the roof. It was actually an attic with the ceiling sloping down to the floor on one side. After Hamburg and London, it had seemed a sanctuary to him. In later years, he recalled going to bed in this room on the night of the accident, listening to the beat of the pump. "I remember well, I was in a very affected state—a fatalistic mood. But then I fell asleep and woke again in the morning to find the danger was over."

They were able to recover most of the radium-beryllium neutron source.

With a neutron gun as powerful as any in the world, the Copenhagen workers found that some materials soaked up neutrons, materials like cadmium and boron, especially. In seeking fresh information on the nature of the uranium nucleus, they were laying a foundation for control systems in future nuclear power stations.

They reviewed their work each week—they called it a colloquium—talked out projects, and debated papers written on the subject elsewhere. In 1935, there came a report from Hans Bethe, then in America, which advanced new ideas on the prospect of capture by the uranium nucleus of a neutron slowed by collision with hydrogen; instantly the talk at the weekly colloquium came to a pause; Bohr ceased his pacing and sat down, his face blank, his eyes glazing. He was so still, so quiet, those around him became alarmed; and then he came to life, a smile spreading over his broad face. He said, "Now, I understand it."

Bohr's imagination and power of thought had brought him to

the second high moment of his theoretical career; the concept of the compound nucleus was born. This was the idea that led to a picture of the uranium nucleus being "like a droplet of water."

He reasoned: a neutron thrusting into this liquid drop, packed as it was with particles held together by mutual attraction, must collide with those other particles, with protons and neutrons already there. The collision energies would be distributed through the many particles, and the nucleus would resonate, quiver, vibrate with this new energy state until one particle or another, or a packet of waves such as gamma rays, was emitted. Nuclear resonance! It was a marvelously precise tool, once it could be measured. With the components of the uranium nucleus packed tightly, the number of collisions would be many millions; but this occurred in such a tiny moment of time as to be almost immeasurable—about one millionth of a billionth of a second! Indeed, reaction to the intake of a neutron in a uranium core meant the water droplet nucleus would quiver, or oscillate, at around 100-million-million-million times a second! This state was, he argued, a period of long duration in atomic time; after this would come the release of gamma rays, electromagnetic wavelengths, from the quivering "droplet."

Bohr's picture of the nucleus was to have an importance far beyond the measurement of resonance; its historic concept was to play a key part in the final solution of the puzzles that emerged from Fermi's bombardment of element 92 in Rome.

In the radio-chemistry laboratory at the Kaiser Wilhelm Institute in Berlin, the Fermi assertions set nerves tingling. And their reaction was their way of science: "Let's see how much of this is fact, and how much is wrong." And so they, too, looked for transuranic matter from neutron attack.

Otto Hahn was now one of the world's great physical chemists. Also, much more was known of the structure of the nuclear citadel since he laid claim to the discovery of radiothorium as a new element. In 1917, with Lise Meitner, he had found that the place in the periodic table immediately below uranium belonged to a new element, which he named protactinium. Now, the vacant spaces in Mendeleev's grouping of the elements were being rapidly occupied. Below protactinium was thorium, with actinium as element 89, and in the place below was Marie Curie's radium, the element that was a key substance in misleading the Berlin thinking.

The same perplexing variety of radioactive products reported from Rome appeared in Hahn's crucibles; and, entranced by the glittering prospect of so many new discoveries waiting to be elucidated by skilled chemistry, Hahn led his associates—he was now joined by Dr. Fritz Strassman—through the maze of bewildering by-products that were to bedevil his hopes and aspirations. He went into what he was to call the "years of tragic confusion".

No sooner had their work of unraveling the Fermi puzzle started than there came a paper written by a former associate chemist in Berlin—Aristid von Grosse, then in the United States—which suggested that one of Fermi's tribe of mysterious elements behaved like protactinium. The challenge was irresistible. This German-born element filled the place below uranium. So—how could uranium be changed to behave like an element of lighter nuclear mass? Such a thought was beyond the realm of physics; yet if it was not protactinium, could it be something beyond uranium? Looking into transuranic country for their answers, they began tracing steps backward with other radioactive debris from neutron bombardment and associating it with radium, isotopes of radium, and thorium, and actinium—not looking for elements much lower *down* the table. Their thinking did not allow for great jumps between elements.

This was the starting post for the contest with Paris, for the checking and rechecking of each other's work. Seeking error, the two teams began moving inexorably toward a momentous discovery.

This year of 1935, when the droplet nucleus was born, which saw the growth of the bewildering family of radioactives from neutron attack on uranium, was the year the Joliot-Curies won the Nobel Prize for their discovery of artificial radiation. And this was the year when Joliot-Curie became prophetic in his Nobel lecture: "We may be entitled to believe that researchers, building up or breaking down elements at will, may be able to bring about nuclear reactions of an explosive nature—veritable chemical chain reactions! . . . we can only look forward with apprehension to the consequences of unleashing such a cataclysm. . . ."

Above all, 1935 was the year when, in America, the final piece, vital to the whole uranium jigsaw, was uncovered.

· 8 ·

Arthur Jeffrey Dempster was Canadian born. After a doctorate at Toronto, he embarked on postgraduate study in Germany, at Munich and at Würzburg University, where Wilhelm Roentgen had found X rays 15 years earlier.

His training in analysis of the chemical elements was intense, and, before World War I began in Europe, he returned to North America to take his Ph.D. at the University of Chicago and to become a naturalized American citizen. From then on his research was based in Chicago. Dempster's talent produced a mass spectrometer, an ingenious hand-built instrument for using magnetic fields to measure the masses—or weights—of atoms and their abundance in various elements.

Dempster assailed the task of reviewing the known elements and their places in the different groupings of the periodic table. Patient, persistent, he added fragments to what was known of atomic structure; he found new isotopes in magnesium and in lithium—and so he worked through the years until he came to the heavyweight, to uranium. Then he made the prime discovery of his life's work. He found an unknown isotope of uranium.

The understanding of the uranium atom was then shaping toward the picture of a nucleus about ten billionths of a centimeter in diameter, holding in its elastic liquid-drop state some 92 protons. These gave the nucleus 92 positive charges, and this positive field was canceled out by exactly equivalent numbers of negative electrons, like planets in orbit around the sun, so that the overall charge of the whole atom was nil. To this model had to be added the neutrons—and uranium had been shown to normally contain 146 of these—making a total particle count of 238. The atom, thus, was known, for its total mass, as uranium 238.

Then came Dempster. His new isotope of uranium was slightly different—it had three less neutrons. The same number of positive charges were in the nucleus—92 protons—the same number of balancing electrons were in the orbiting shell, and since these determine the chemical behavior of the element, the existence of this different atom was only revealed by its different weight. Three particles less did more than make uranium 235 marginally lighter: it made it more unstable, more vulnerable to disruption from intruding neutrons; it made uranium deadly. It was very rare. It came to be found as existing only seven times among each thousand uranium atoms. It was shown, later, to have a half-life of 700 million years—the period when its radiation diminished by one half—as against a half-life of 4.5 billion years for uranium 238. This meant that when the earth was young, uranium 235 comprised about one third of all uranium on the planet. Scientists were subsequently to speculate that had mankind developed earlier in the history of the planet, atomic energy would have been easier to discover and cheaper to run, while if evolution of man had been much later the atomic age might never have dawned.

The world should have taken note of this modest, little-known man, and what he had found; but very few people realized what he had discovered. Uranium 235 was just another number, another isotope among a bewildering increase in isotopes being culled from neutron attack on the uranium nucleus. And it was "much too rare to matter." For that very reason, Otto Hahn and Lise Meitner disregarded it; it did not rate investigation among the plethora of what they took to be isotopes of radium, actinium, and protactinium.

What they could not fit around the top elements, as radioactive isotopes, they transported to their "transurania" concept. Otto Hahn was to say in retrospect:

> Admittedly, it was faulty reasoning. We thought them to be transuranic elements with atomic numbers higher than 92; everyone who worked on the problem reached the same conclusion. There seemed no other ... we showed von Grosse was wrong, that Fermi's isotope with a 13-minute half-life was definitely not protactinium. We decided Fermi was right, that this was one step beyond uranium, an element with a nuclear number of 93. We called it eka-rhenium.

They were convinced the new matter was a kind of repeat in higher structures of the group of elements surrounding platinum, which was element 78 on the Periodic Table; rhenium was 75, and *eka* was prefixed because they believed the new substance would have a similar behavior. In the same way they came to label other neutron debris as eka-osmium, eka-platinum. And, because they used barium as a carrier chemical in their separation processes—and this has a similar chemical behavior to radium—they labeled some of the irradiated products as radium isotopes.

The mistake made in Rome was compounded. "Suspicion should have been aroused," Hahn recalled, "but the situation was so complicated, strange facts were accepted, and many carefully conducted investigations led to wrong conclusions. Nobody could be blamed. There seemed no other choice at the time."

Transuranic elements were there, however, masked and hidden, and in such small amounts their radio signals were blotted out by the activity of a nuclear reaction not to dawn on human minds for another four years.

As well, Hahn and his colleagues were under growing tension, mounting oppression. While Hahn was abroad, lecturing in Ithaca, New York, his Institute colleague, Fritz Haber, had declared he did not want to be treated any differently than his Jewish colleagues. Haber was broken, soon to die in Switzerland. Hahn found on his return that the regime he thought "would not last long" had an iron hand that slowly tightened around the formerly closed-in world of his laboratory and that his years of "tragic confusion" were to be burdened by suppression. Members of the Nazi Party worked in the Kaiser Wilhelm Institute and individuality, talent, skill, and dedication counted for nothing. Still they pressed on along their twisting, torturous chemical path, seeking their transurania.

· 9 ·

In Copenhagen the hunt with the neutron was far less grim than in Berlin. For Otto Frisch these were happy days. He worked with another Viennese exile, Georg Placzek.

In one of their experiments they needed thick layers of solid gold to test Bohr's theories, and Placzek had the idea of using several Nobel Prize gold medals given to Bohr for safekeeping by laureates in Germany who feared confiscation for the Nazi war machine. Frisch and Placzek found gold to be a strong absorber of neutrons, that the resonance factor was thousands of times below what was expected. Frisch noted, drily, "It was a great satisfaction that these otherwise useless medals should serve a real scientific purpose."

Later, when the Germans overran Denmark these same medals were again saved from confiscation; Georg Hevésy dissolved them in nitric acid kept in an old beer bottle during the war. Afterward, the gold was reformed and the medals again struck for the original owners.

And for Frisch there was another wry satisfaction. Bohr decided to add to the Institute's armament by installing a cyclotron—the first outside America—but in the old-style building neither doors nor windows would allow passage of the 40-ton magnet, and a wall had to be breached. Then, when a million-volt generator was needed, Frisch traveled to Leipzig to deal with the German makers; he found the place was like a war camp, with hatred for Jews openly displayed. He noted sardonically, "They treated me like an honored and welcome guest. Those same people who had sacked me under Hitler's racial laws, were quite willing to forget race when money entered the picture."

In the Danish capital, the hunt for the secrets of uranium was intellectual; there were fine companions—and there was Bohr. After work, some evenings were spent at the great man's home at Carlsberg. Frisch remembers, "Some of us literally sat at his feet on those happy nights. We listened to his words, and I felt at times that here was a Socrates come back to earth to help us be wiser than we were. He tossed out challenges in his gentle way. . . ."

Cycling home through the streets of Copenhagen, wet with rain, or reeking with the scent of lilac, Frisch would return to his room at the Institute, to his books and his music, and to reflect. As his aunt was to remark, it was grand to be alive in a world where there were people like Niels Bohr.

It couldn't last much longer; this fragile illusion of pure science, the international camaraderie among the uranium researchers.

· 10

Death and discovery brought drastic change in Paris. With the death in mid-1935 of Marie Curie, from radiation effects, an era of glory in French science ended. However, Irène and Frédéric Joliot-Curie were in firm control at the Radium Institute, and the family quickly extended its sphere of influence; Joliot-Curie was elevated to a tutorial seat at the Sorbonne and, the following year, became professor of physics at the Collège de France. Like Fermi in Rome, he turned away from the prospect of painstaking inquiry into the nature of the deceptive fragments of neutron attack on uranium; like Bohr in Copenhagen, Chadwick in Liverpool, and others in Europe, he was drawn to the prospect of engineering as a tool for probing the uranium nucleus. He spent much time developing the cyclotron to a 7-million electron volt capacity, as well as installing banks of cloud chambers and other expensive and heavy equipment.

Irène at once filled her mother's shoes at the Institute. She always had position and influence as the gifted daughter of "la Patronne"; now, she had status as a Nobel laureate, and the powerful position of a leading world worker in the field of radioactive substances. She never forgot her background and was ever conscious of her family name and the nobility and selflessness of her parents with their utter disregard for financial reward.

So it was that when queries on her reported results, sharp questioning of her accuracy, came from the rivals in Berlin, she saw them as spiteful and envious harrying rather than honest critiques from international colleagues.

It did not make sense for a lone worker to tackle the same broad sweep being made in Berlin; so she decided to isolate one of the more interesting of the active by-products reported in the Fermi papers.

The Shinkolobwe radium, presented to her mother by the grateful Union Minière Company for her services at Oolen, was lifted from its recess and its energetic stream of alpha particles directed to beryllium to strike neutrons on to a uranium target. Precise and painstaking, Irène watched the process for many hours; then the uranium went into the separation, precipitation processes her mother had shown her. Irène now felt the need of a skilled physicist with a flair for measurement techniques; she invited a young Yugoslav researcher, Paul Savitch, to join the project.

They isolated the substance Irène sought and established that it reduced its radiation by one-half each three and a half hours. It gave her time to work on the chemistry, but she needed much more. Some deep instinct moved her at this stage of her project; she felt the claims of transuranic elements to be wrong. Fermi, Hahn, Meitner—all of them were looking upward in the scale, thinking the neutrons created some element heavier than uranium. What if the encounter with each nucleus resulted in a loss of more energy than was taken in—a loss of energy, which meant a loss of mass, which meant a change to an element in a lower position in the periodic table?

This 3.5-hour half-life substance was a puzzle. Its chemistry showed no definite characteristics that would tie it in as an isotope of a known element, but it was certainly not a transuranic substance. She was convinced of that. The experiment was done over and over again, and the nearest correlation she could make was that the material behaved somewhat like element 90, thorium. They decided to write a note on the work and circulate it for consideration. Irène's uncertainty was reflected in her phrasing; still she suggested it might be a form of thorium—and that opinion sounded as sharp a note in Hahn's laboratory as did the proposal that one of Fermi's by-products might be a version of protactinium.

In October, 1937, the Italian authorities invited a number of leading European atomic scientists to celebrate the centenary of the death of Galvani, inventor of the electricity storage battery. Among those invited were Joliot-Curie, Hahn, Max Planck, Bohr, and Lord Rutherford; however, for the first time in his life, Rutherford was ill and his place was taken by Professor Mark Oliphant, who then held the chair of physics at Britain's Birmingham University.

The Galvani anniversary celebrations were marked by two important developments. While there, Oliphant was handed a cable telling of Rutherford's sudden death; he flew back to London and drove to Cambridge in time to see Rutherford's corpse before the remains were cremated and the ashes placed in a corner of Westminster Abbey, next to the grave of Isaac Newton.

The other event at the Italian celebrations was the first meeting between Frédéric Joliot-Curie and Otto Hahn. The short Prussian chemist spoke warmly: "I have seen much of what you have done with Madame's radium, and I feel I know you as a friend." Then he referred to Irène's latest work: "I also feel great friendship and have admiration for your wife's work. Nevertheless, my friend, I think she is very wrong about thorium. I have decided to repeat her experiment, and I think I shall soon be able to show she has made a mistake."

Once back in the Kaiser Wilhelm Institute, Hahn soon found time to join with Lise Meitner to show how the Frenchwoman had been in error. They found the 3.5-hour substance, isolated it, and proved conclusively that it could not be thorium. They wrote Irène a letter to say this; but they could not say what the active substance was. It was a mystery to them, one of many mysteries; perhaps it was some new matter beyond element 92? But it could *not* be thorium.

In the room her mother had used at the laboratories in the Rue Pierre Curie, Irène suffered her rejection in solitude. The chemical evidence from Berlin had to be accepted. She had to agree that her inference that this matter might be thorium was a mistake; the suggestion should not have been made without strong evidence. And, tragically, she was about to repeat that same mistake in her drive to show something new.

In the laboratory with Savitch, she reconsidered her quarry. If this was not thorium, it was *most* certainly born of uranium; and the instinct to look away from transuranic matter was strong. If it was born in element 92, and it was not protactinium, or thorium—could it be actinium, element 89?

All her skill and knowledge were spread over week after week on this single objective. However, she never could show this 3.5-hour matter to be anywhere close to the starting point of uranium. What was it really like, as a chemical, apart from being a radioactive body?

She made exhaustive, eliminating tests. The weeks were passed in showing what the substance was not—never finding what the substance was; then, when using lanthanum as a carrier material from one of her solutions, she saw it behaved like lanthanum! How could that be? Lanthanum was giant steps away in the nuclear world; lanthanum was element 57, almost halfway back down the periodic table! How in heaven could a middle-order element suddenly appear in this purified uranium, in element 92?

At this crucial moment of her career, her mind, her imagination, could not make the leap. The puzzle remained unsolved. Her 3.5-hour substance was not conclusively identified. She expressed her bafflement in cautious language in her report—"the substance is *similar* to lanthanum. . . ."

A certain second Nobel award slipped through hesitant fingers, away to her rivals in Berlin for whom it was now almost a way of life to show how the Frenchwoman was wrong, how she was not as good as her mother.

· 11 ·

Anschluss! Austria was annexed into Hitler's Greater Reich. The puppet chancellor, Seyss-Inquart, went to the microphone that fateful Sunday, March 13, 1938, to do Hitler's bidding. Austria was no more; it was *Ostmark,* a region of the German Union.

By this single act, all Austrians of Jewish birth were caught in the net of Nazi hatred of the *Jude* and were subject to the racial decrees and the persecution they generated; and thus, Hitler lost yet another chance to be first with the atom bomb.

From this day forward Lise Meitner found her position growing steadily more parlous. She continued her work with Hahn and Strassmann at Dahlem until the summer, quietly operating in the background in an effort to escape notice—in vain. Soon, Nazi officials overseeing the Institute staff ordered her to wear the Yellow Star of David at all times.

The Yellow Star was a badge inviting discrimination—all across Greater Germany. Her colleagues of long years in the Institute could shut their eyes to it; but when she went home at night through the streets she was insulted, jostled, and at one time punched by raucous uniformed youths.

Through the weeks of growing danger she stuck with her task, working with Hahn and Strassmann on the nature of an elusive product of neutron bombardment with a half-life of 60 days. Hahn made protest after protest, on her behalf, to no avail; she was always and ever "less than fifty percent Aryan."

In Copenhagen, her nephew Otto Frisch related his anxiety for the safety of his parents and his aunt to Bohr, who was both consoling and helpful: "If she can get a visa and come out, there are dozens of places that would welcome her with open arms."

In Berlin, Hahn and Max Planck asked the President of the

Institute, Privy Councillor Bosch, to plead for a visa for Lise. The threat against her overrode her status in research. Days later, they were called back to his office; Bosch was sad, very solemn. Hahn and Planck heard the words read in reply to Bosch's entreaty—a letter of refusal from the Nazi education minister. Lise Meitner was trapped; nobody could help her—except Bohr.

A professor wrote to Holland from Zurich, and three Dutch professors went to The Hague to make certain arrangements.

At the Kaiser Wilhelm Institute, Lise wore the Yellow Star and worked on the 60-day substance suspected of being something akin to the platinum group. They called it eka-iridium and wrote their paper for the editor of *Die Naturwissenschaften.* It was posted on July 10, some days later. It was their last joint report from more than 30 years of cooperative research.

Lise then went on holiday like an innocent tourist and arrived with a small bag of possessions at the Dutch border. Unlike an innocent tourist she was passed into Holland—without question—without a visa. Soon she was united with her nephew in Copenhagen and, a little later, accepted the offer of a post in Stockholm. It was before she left Denmark that she remarked to Frisch, "It is good to live in a world where there are men like Niels Bohr."

(Hitler had now driven from Germany two people whose brains were to unfold the means by which he could have made himself world dictator.)

Otto Hahn missed her badly. Her mathematics, her theory, her acute sense of the inner physical world left a gap in his laboratory. Through what was left of that summer—with Europe hanging on the brink of war over Hitler's claims to Czechoslovakia's Sudetenland—Hahn pursued his course of trying to identify the by-products of uranium bombardment.

His remaining colleague, Fritz Strassmann, was not directly molested or restricted by the Nazi officials. Somehow, they overlooked him until he applied to the University of Berlin to be appointed assistant professor, saying he wished to enter academic life. He was told he would not be considered unless he joined the National Socialist Party; this was according to the laws applying to higher education establishments. Strassmann refused and remained at the Kaiser Wilhelm Institute, where such laws did not apply. Thus, he was on hand to help Hahn try to show that Irène

Joliot-Curie was, again, in error with her suggestion that the 3.5-hour radioactive substance she had found, acted like lanthanum.

Meanwhile, Hahn wrote to Lise Meitner in Stockholm with the details of a paper he proposed on the likelihood of some of his new radioactive fragments being isotopes of radium. She replied at once, "I implore you not to publish that incomprehensible result—not until you are very certain it is correct."

Hahn was left hovering between a jump up the scale from element 92 or a jump downward, and, in the pause of uncertainty, turned his attention to Irène's suggestion that one of the fragments behaved as though it was lanthanum. He asserted, later, "Since we believed it to be a transuranic substance, with unusual characteristics, we decided on a thorough investigation." The Hahn techniques went into operation.

There were illusions of peace from the surrender to Hitler by France and Britain at the Munich conference.

But news of inhuman treatment of Germany's Jews sent a shock wave across Europe. Hatred was bound to be met with hatred. On November 7, 1938, a young man, almost demented at the utter cruelty of Nazi deportation of "stateless" German Jews to Poland, shot down Ernst Von Rath, Secretary of the German Embassy in France—a noted Jew baiter. Savage vengeance was unleashed throughout the enlarged Third Reich. Jewish community leaders were made special targets for roaming gangs of brutal storm troopers; teachers, traders, artists, musicians, academics were attacked.

In Vienna, the night of Wednesday, November 9, 1938, Nazi gangs burned a hundred synagogues, smashed the windows of shops and stores and looted the contents. *Kristallnacht* opened a week when the *Judenmaterial* were hunted like animals, when sentences were passed without trial, when thousands were herded into the new concentration camps.

Among the victims in Vienna was the father of Otto Frisch.

Otto Hahn was a world leader in radiochemistry. He had framed rules over the years for the precipitation of minute fractions of matter, for winning mere traces of radioactive materials from uranium solutions. His teaching was even then being followed by young graduates in other lands. In California, on the

Berkeley campus, a serious-minded, gangling young researcher from Ishpeming, Michigan, who had taken his doctorate only the previous year at the age of 25, read all that Hahn had written and hailed him as the "father of radiochemistry." His name was Glenn Theodore Seaborg.

And—in the late months of 1938—all Hahn knew went into the assault on the veracity of Irène Joliot-Curie's cautious lanthanum suggestion. The 3.5-hour material was plucked from the Berlin solution; then they used analytical precipitation with lanthanum as the substance that would "carry" the object for identity. They also used barium—next to lanthanum as element 56—and zirconium, element 40. It was a net to indicate a group the quarry might fall into, and it came out with a number of different radioactive substances.

"It is a mixture," Hahn concluded. "It is made of several substances, which is why it is difficult to determine."

Yet even though they found the tests with barium intriguing, Hahn still clung to his notion of radium isotopes.

He started again, with ten grams of a uranium preparation, and when he used barium as a precipitation agent, he was again nonplussed. He recalled, "We could not immediately distinguish the barium from radium isotopes, but, since *barium from neutron impact on uranium was simply out of the question,* the substance could only be radium—element 88." [Author's italics]

He then reached a conclusion that was, to him, a violation of all previous experimental evidence in nuclear physics. He made the suggestion that his radium isotopes really were barium and that what he had taken to be actinium isotopes possibly were lanthanum after all. He entered into his report the same hesitancy that had marked Irène Joliot-Curie's result.

He and Strassmann wrote a joint paper—"with reluctance, because of the extraordinary results." And they used a sentence that was to become famous in science: "Speaking as chemists, we even have to say that these new substances *are* barium, not radium . . . ."

Just as Irène had halted at taking the great leap, so did Hahn and Strassmann. Most of their report had been compiled as they went along; in it, they had referred to "radium isotopes." Now it had to be changed. It was near the end of the year, and they wanted to publish their work quickly. They altered their words, still hesitating, from "radium isotopes" to a vague "alkaline earth metals" and *not* to "barium isotopes."

Facing this challenge, convinced of their chemical results, but unable to come to terms with the contradiction of known physical laws, Hahn sat at his desk in the quiet of evening and confided his situation by letter to Lise Meitner. He enclosed a copy of the paper they would publish and expressed his perplexity. It was posted to an address in Sweden where Lise Meitner was spending Christmas.

Still brooding over the puzzle, Hahn went to his home to enjoy the holiday. It was the last peaceful Christmas Berlin would know for six years; it marked the last fundamental discovery in uranium in Germany.

· 12 ·

Otto Frisch crossed the Ore Sound to Sweden to spend Christmas with his aunt. A stabbing wind blew off the near-frozen Baltic as he walked the deck of the ferry, crowded with holidaymakers. He was in no such mood; the plight of his parents caused him serious anxiety. He had been moved to speak to Niels Bohr of his worry and had been advised, "Enjoy your holiday. Give your aunt my warm regards—and do not fret too much about your father. We'll see if we can pull some strings."

But Frisch left Copenhagen with no false hopes. News had just come to the Institute that even Enrico Fermi had been forced to desert his homeland. He and his Jewish wife and children had attended the awarding of the Nobel Prizes in Stockholm in December and had gone to the United States; so what illusions could he hold on the gates of a concentration camp opening to release his father—because Niels Bohr pulled a few strings?

The thought of his mother's despair worried him all through the long train journey to Kungälv, a resort overlooking the Skaggerak, with the coast of Norway 150 miles distant. When Frisch joined his aunt for breakfast she was intently reading a letter. He greeted her, sat down, and started to speak, "I want to discuss with you an experiment I am planning, and—"

Her face was creased in concentration as she read. Then she

looked up, puzzled: "Otto, you must read this news from Hahn; you *must* read it!"

Frisch would always remember being unwilling to read Hahn's report; he wanted to talk about his own uranium experiment. But Lise Meitner was very firm; she thrust the letter into his hands, insisting. He read as he ate, and he found the suggestions to be so much at variance with the accepted ethos in physics that he was at once skeptical.

After breakfast, they walked out of the hotel together. Frisch put on skis, but his aunt wore strong boots; she said she could get along just as quickly that way through the heavy snow—and she proved it. She kept up with him and, all the while, argued the veracity of Hahn's results. When they were in the woods, she halted him with a restraining hand.

"You can't doubt Otto's words! I've worked with him for thirty years. He's one of the finest chemists on earth. If he says the barium is there—it *is* there! There is some other explanation we have not yet found."

They sat together on a fallen log between the bare trees, and there, among the snows of Sweden, the understanding of uranium neutron disintegration was achieved.

No larger fragments than alpha particles or protons had ever been known to emerge from the uranium nucleus. They dismissed the idea that numbers of these fragments could be "chipped" from the atom's core because they knew of no energy powerful enough to do that.

Nor could it be that the nucleus could be sliced in half. Why? Because it was not brittle, not solid, like metal. It was, as Bohr had stressed, just like a liquid drop, though very, very dense.

And what, they reasoned, if by some means that liquid drop could divide—what if it could change its shape, become elongated? What if it could become *two* drops? Without being severed by force, or ripped apart; if it did it slowly? Of course, there were powerful forces to react against such a process. They knew that. Surface tension operated against water drops, against beads of liquid mercury. But uranium nuclei were electrically charged. Each of the 92 protons was a unit charge, and wouldn't this reduce the surface tension effect?

There were scraps of paper in Lise's handbag, some in Frisch's pocket. They sat on the log and calculated from known forces.

Indeed, it was so: The charge on the uranium nucleus was large enough to destroy the surface tension—almost entirely—and that they deduced meant that the tightly packed uranium core was very unstable.

"It is a very wobbly droplet," Frisch proposed, "so unstable that it is ready to split itself in two—under provocation."

And provocation was the impact of the arrival of another neutron. However, it wasn't as simple as that; there was yet another problem.

Suppose the single uranium nucleus became two parts—the atomic number of 92 was roughly halved and you had, as a result, an element around barium, element 56, or Irène Joliot-Curie's lanthanum, element 57? What else would be there? The other half of the uranium nucleus would be either bromine, element 25, or krypton, element 36; and krypton was a gas. And following separation, surely, these two new atoms would have to be driven apart by their mutual electric repulsion to maintain their new identities. It needed, they reckoned, some 200 million electron volts of energy for this to happen. Where would that great energy come from? Lise Meitner computed from the known formula on the packed mass of the uranium core that division of the nucleus would mean a loss of mass. She calculated this to be equal to one-fifth the mass of a single proton: "And by Einstein's equivalence—by his formula of energy equalling the mass multiplied by the speed of light squared—one-fifth of the mass of a proton is just about the required 200 million electron volts!"

It fitted precisely into their speculation; Frisch had to remind himself in the days that followed that it was just *that* at present— speculation—no matter how excited he felt. It was still an idea, maybe a new law on the behavior of matter for which they did not, as yet, have a name.

Back in Copenhagen, Frisch walked into the Institute building bursting to blurt out his news—and the person to hear it first would be the world's greatest living theorist, Niels Bohr. He went quickly to his attic room, dumped his bag, washed his face and hands, and went looking for the Director. Bohr was giving last-minute instructions before leaving for America; he was shortly to depart to catch the liner *Drottingholm,* to be in New York in good time to spend a few days at Princeton University before going to the Fifth Washington Conference on Theoretical Physics. It was not his nature to brush aside the unexpected caller; he smiled, extended a warm greeting, and asked after Lise Meitner.

He listened entranced as Frisch related the historic concept. On hearing of the brilliant, classical detective work, the same misty look came into his eyes as on the day the liquid-droplet nucleus was conceived. Again he was suddenly vitalized, his eyes aglow. He lifted his hand and, with the palm flat, smacked it hard into his forehead in a gesture of self-rebuke and exclaimed, "My God! How wonderful! What idiots we've been not to have seen this before!"

In a few seconds he had worked out the process—the quivering liquid drop, disturbance from the new neutron, and then, in a flash of atomic time, elongation, division into two parts—release of energy to drive them apart: the creation of two new middle-order atoms.

A century and a half after Klaproth had prised uranium from the Bohemian ores, it seemed the secret of its energy was, at last, unfolded. But, not quite; not yet! The final picture was to be revealed in the weeks ahead, from a reunion at Princeton between Bohr and a former student.

Before he left, Bohr shook hands warmly with Frisch: "This is just as it should be—marvelous, wonderful! Are you and Lise writing a paper on it?"

Frisch replied there had not, as yet, been time; and they needed to get his reaction first. Bohr said, "Write your paper immediately. I will say nothing to anyone until you are in print with this great discovery." Then he was gone, racing for his ship with his secretary and assistant, Leon Rosenfeld.

Frisch felt alone and a little at a loss as to what to do. It would have to be a joint paper, of course, with his aunt. It meant a series of telephone calls to confer on what to say, and that would take time; also, he needed to talk with others, to compose his thoughts on the approach, on how to describe the mechanism of the division of the uranium core. He and his aunt had pictured it like a droplet of water—as though it was a living cell, an organic entity, separating into two biological units. He remembered there was an American biologist working with Georg Hevésy; Frisch went to him and asked, "What is the term you use for the process of a single cell elongating, and separating into two cells?"

The biologist, William A. Arnold, provided the name for the

critical process in the liberation of atomic energy. "Fission," he said. "We call that reaction—fission."

Frisch called Stockholm and told his aunt he proposed to use the term "fission" to describe their explanation for the mystifying results Hahn had obtained in Berlin. He then discussed, with Georg Placzek, the paper he planned to write.

"Look," said Placzek, "I find it hard to accept. First, you tell me how the uranium atom suffers from the disability of instability, and that's why it kicks out alpha particles. And now you tell me it suffers from this additional affliction! It seems very unlikely."

He went on, "I think it's like a man hit by a brick falling from a building, and when you take his corpse to the hospital you find he died from cancer. It's too much of a coincidence. I don't like it! Can't you do some experiment or other—like looking for the fragments in a cloud chamber?"

This response brought Frisch up sharply.

"Oddly enough," he said, "I hadn't thought of looking for proof, for experimental evidence."

The days were quickly slipping by. Bohr had left on January 6; composing the letter with his aunt, talking over and framing his paper had taken five days, and now he would have more delay in setting up an experiment. What was the urgency? Apart from Lise Meitner and Placzek, only Bohr knew of the new theory, and he had undertaken to say nothing until their report was in print and credit placed where it belonged. As well, Frisch's tardiness was strengthened by his natural reticence and modesty. He told, much later, how his state of mind masked the need for urgent action.

"I, personally, had no claim to intelligence or originality. I was just lucky to be there when Lise Meitner received that classic letter from Hahn; and, anyway, at first I didn't want to be bothered reading their report. Then I had to be nudged into making an experiment to give some empirical proof to ideas in a paper that was almost completely written. So—I hadn't much claim to urgency!"

Preparations for his experiment took more time than expected. He had no access to a cloud chamber in which to show and photograph the tracks of fragments from fission of uranium;

he had to enlist the Institute's glass worker to help build a special ionization chamber. The Bohr birthday radium-beryllium source was in the well in the cellar and had to be installed in his laboratory where its neutron beam would strike a sample of uranium oxide.

First, he estimated the firing range of the fission fragments would be a matter of centimeters, and from this, he determined the size of his ionization chamber. These flying bits of matter would emerge from a layer of uranium and ionize the gas atoms in the chamber. The gas would then be conductive to electricity and provide an instrument of measurement, of range and intensity.

He built a small chamber with a narrow glass ring joined by two metal plates on one of which was a layer of uranium paste. It had to be cunningly rigged and monitored constantly. Once he had the pieces, it took a whole day to assemble and prepare for fission measurement. It was late on Thursday, January 12—after 4 P.M.—when he finally sat down to start the tests. It was the most intensive lone study he had ever made—and the most momentous. Hour after hour he kept watch, recording all the time, copying down rows and rows of figures—the first record ever made of uranium fission.

Frisch recalled, "I willed myself to keep on working. The rate of particle omission was low, and I had to make sure of my results because I knew every other physicist in the world would follow up, and I had to be accurate. I had the Geiger counter going and a meter ticking over—just like a waiting taxi!"

He spaced the records apart in time. The neutron source was striking the uranium layer; the resulting bursts were registered in the flow of charge across the chamber. Two minutes, ten minutes; take away the neutron source, and then take control readings. Start again, establishing the range of the fission particles.

Strangely, he did not feel excitement or emotion, only satisfaction. He mused, "Well, this is fine. Now, when I say this thing is true, I can show that it's true."

He was convinced. Here was direct proof. He wearily climbed the staircase to his garret and noted it was now 5 A.M. on Friday, January 13, 1939. Two hours later his slumber was shattered by a roar of sound that raised alarm in his half-awake mind. Somebody was hammering on the door; a voice was calling him, urgently.

The Institute janitor stood there, anxious, troubled, and holding an envelope against his apron,

"I called, and called—I couldn't wake you! They brought this cable for you and said it was important."

Frisch sat on the edge of his bed and opened the cablegram. It was from his mother in Vienna. At first, his tired eyes and sleep-fogged brain could not adjust to the words.

His father had been released by the Nazis. It was a conditional release; his parents could leave Greater Germany—but without possessions, relinquishing all their money and property to the Reich. They were leaving in two days to join Lise Meitner in Sweden. Bohr's discreet "strings" had brought results.

Frisch could no longer deny emotion. He tried to sleep but could not. He went to the laboratory soon after breakfast and talked with Placzek; others came into the room as he expounded the results of his fission tests. One said, "Should we not send a cable to Bohr?"

Frisch still felt tired, drained. He had to finish the paper and decided to concentrate on the writing and not bother with expensive telegrams. It took him until the Monday, January 16, before he at last signed the paper, in which was the phrase "of present ideas about the behavior of heavy nuclei, an entirely different and essentially classical picture of these new disintegration processes suggests itself."

He wrote it in English; he sent it to the British magazine *Nature,* in London, because it was reputed to publish important new work quicker than any other journal; also English was the usual "lingua franca" of the Institute.

Some hours after the Frisch letter dropped into the postbox in Copenhagen, the *Drottingholm* came out of the heaving Atlantic into New York harbor. Niels Bohr had been untroubled by the heavy crossing; each day he had walked with Leon Rosenfeld, talking excitedly of the marvelous findings Frisch had brought back from his holiday with Lise Meitner. Incessantly Bohr had gone over and over the evidence, "this beautiful piece of scientific deduction." He had gnawed like a terrier, attacking the concept from all angles, growing more thrilled with its physical and theoretical symmetry. He gave Rosenfeld little relief from the subject—and completely forgot to warn that the idea was still confidential, that he had undertaken to keep it secret until the Frisch-Meitner report was in print.

Old friends were waiting at the dock to greet the great physicist. Enrico Fermi, his wife Laura—they had been in America only two weeks—were there with university leaders and former students, and among these was a young man who had worked in Copenhagen some years before, John Archibald Wheeler, aged 27 and assistant professor of physics at Princeton University.

As soon as Bohr and Rosenfeld were through customs and immigration, Wheeler wanted to carry both of them off to the university; Bohr declined. He had business in New York which would take a day or two, he said, and he would then rejoin his companion, Rosenfeld, at Princeton.

The scene was now set for unwitting disclosure. The warm, friendly circle at Princeton enclosed Rosenfeld. He was invited to the usual meeting of what was known in the university as the "journal club," held in the Palmer Laboratory. He was welcomed and quizzed politely for any news from Europe—and Rosenfeld imprudently obliged. He knew nothing of Bohr's undertaking; he was under the impression the Frisch-Meitner report was in print, or imminent. And he told the tale as he had heard it from Bohr and added Bohr's own analysis sketched in the many hours crossing the Atlantic. Rosenfeld laid the greatest—as yet unpublished—discovery in atomic science at the feet of the grateful Americans.

In the course of the critical years of 1939 to 1941, three priceless seeds of information were brought from the Old World to the New World, to flower and flourish into collective and effective expansion: three seeds, for which little public recognition was given, that were of tremendous importance to the defeat of America's enemies. They were penicillin, radar, and uranium fission.

Excited at the news, the Princeton group had no reason to remain silent. John Wheeler at once organized a meeting of leading physicists. When Bohr arrived on the scene, days later, he found an irretrievable situation.

To protect the discovery rights of Frisch and Meitner, he at once set to work, with Rosenfeld, to write an article of about 600 words, explaining how they had elucidated the puzzle which had baffled world physicists for half a decade. He also had to face the meeting Wheeler had called; he told them the facts of the new phenomenon and asked them to observe that they all knew about this marvelous piece of pure science through the kindness of Lise Meitner and Otto Frisch, whose work was yet to appear in print.

The situation did not hold. On the eve of the Fifth Conference on Theoretical Physics—to be held at George Washington University, in the District of Columbia, on Thursday, January 26, 1939—an excited and breathless Dr. Willis Lamb burst into the office, at Columbia University, of Enrico Fermi with the news Rosenfeld had leaked, which Bohr had now confirmed. Fermi sat stunned for some minutes, thinking back to the variety of indications the element had given in Rome, in 1934. . . .

In Washington, the editor of *Science Service* received a copy of *Naturwissenschaften* the day before the physics conference opened. He could read German; he sent Robert Potter to quiz Niels Bohr: "Sir, we have this article in German by Hahn and Strassmann which makes a strange claim about barium from neutron attack on uranium. Can you comment, please?"

Chance had thrown open a gate to release Bohr from his straitjacket. For a week he had been sending cables to Frisch— "Act quickly on paper," and "Urgent you substantiate with experiment." And he had had no reply. In Copenhagen's quiet serenity this urgency had seemed unreal; the experiment was done, the paper had gone, and there was nothing more to consider. Until his son wrote him, Bohr did not know the work was completed.

Nature did not publish the paper until February 11. With Potter's question, Bohr's release had come.

A paper had just been read at the conference; the chairman was calling for comments or questions. The Danish genius stood up and walked to the rostrum:

"I would like to make an announcement—of special importance. . . ."

The American uranium saga had opened.

· 13 ·

Dr. Glenn Seaborg strolled to the weekly seminar held by Professor Ernest Lawrence in the physics department on the Berkeley campus, unaware of the excitement sweeping the American world of physics.

Seaborg was a chemist at the University of California and then was research assistant to Gilbert Newton Lewis. He had won his doctorate in 1937, after some years of economic struggle living in a shared room in a boarding house on around $12 a week. All this time he had been a disciple of Otto Hahn. "His book, which set out laws for precipitation of micro-amounts of radioactive materials in solutions, was always my bible."

It was because of Hahn's influence that Seaborg moved from pure chemistry to the world of particles, neutron behavior, and nuclear chemistry, which was within Lawrence's purview. The logic of the radiochemical world always moved him to excitement. He knew intimately of the Rome work, of Fermi, Segré, and d'Agostino, and it was in the reading of their works that his mind "first got excited at the prospect of breaking the element barrier beyond uranium, of getting higher in the scale than element 92." That burning interest had linked his chemistry closely with the physics in Lawrence's faculty in the Le Conte building.

He found the seminar abuzz with excitement. As usual, Lawrence was presiding. Robert Oppenheimer was there and Edwin McMillan, who had a room next to Seaborg and who was sitting with Segré, of the original Fermi team. There, also, was Luis Alvarez, looking rather odd with half his hair cut! Apparently Alvarez had been in the chair at the Student Union barbershop reading a newspaper when, suddenly, he let out a yell, leapt from the chair, pushed aside the startled barber and, waving the paper, the barber's white sheet still flowing from his neck, had run to the physics department to spread the news: The *Science Service* news organization had syndicated Potter's story, "New Hope for Releasing Enormous Stores of Energy Within the Atom."

The New York Times had trumpeted: "Vast Energy Freed by Uranium Atom"; but the paper's report appeared on an inside page. Other editors were even more cautious; nevertheless, the story of Bohr's news had wide coverage. Some at this Berkeley seminar had heard direct by telephone from the Washington conference; some were openly skeptical, believing the details had been garbled in transmission—but not the man whose depth of experience and background Seaborg esteemed.

Across the table Julius Robert Oppenheimer was filled with excitement. A slender, tense young professor of physics—then only 35—he was already recognized as the natural leader of the new American school of theoretical physics. He had a background of study at the Cavendish with Rutherford, in Copenhagen with Bohr, with Wolfgang Pauli in Zurich, and had won his Ph.D. under Max Born in Göttingen—at the tender age of 23 years. The son of Jewish immigrants from Germany, he had a cultural charisma that brought him high regard from students and faculty members alike—and, though he could claim no fundamental discovery, he had published many papers and was recognized as a popular personality, a trait that was to ensure his place in the development of this new concept of the behavior of matter.

Oppenheimer was not an experimental physicist—his work had been in the rare air of abstract science—and though he was not a chemist like Seaborg, the explanation brought by Bohr of the barium found in the Berlin experiment was a lightning flash in his mind—totally acceptable. So he spoke freely of the implications for ideas on the relationship between the elements.

In Seaborg's mind the revelation struck "like a thunderbolt." Brilliant! Thrilling! The shadowy forms of the puzzles of the past now had shape; processes which had mystified were suddenly vivid, clear, and simple—and disturbing.

Seaborg unfurled his long legs and left Le Conte Hall, went out into the night to roam the cold, wet streets until the early hours; and, with his surging thoughts on the new nuclear situation came chagrin:

"I was disgusted with myself that I had not seen this explanation. It seemed so beautifully obvious. I had had all the clues. Alvarez, Oppenheimer, Segré—we'd all had the same clues, and we had failed. Like Bohr, each of us was saying—how could we have missed this?"

What now was left?

Those elusive transuranic elements Seaborg had sought, after Fermi had set the trail *beyond* uranium—were they gone, were they illusory, atomic ghosts dissolved in the disclosure of the rupturing uranium nucleus? Did this new leap in understanding mean they could not exist? Were all the things in Hahn's crucibles middle-order isotopes? Wasn't it *now* that the search could begin for the *real* elements beyond uranium—to see whether they existed in nature, whether they could be created in the laboratory? They could exist; they could be the building bricks of materials such as man had never known.

No, fission was *not* the end of the transuranic search for Seaborg. Far from closing down a path, fission had opened a new approach into the world beyond element 92.

The coming of fission to the United States was an unparalleled stimulus—in pure science. Soon, however, the minds of exiled scientists who had suffered from abuse of power feared the misuse of uranium by the very country from which came the chemical clue to fission.

Only two days before Niels Bohr had left Denmark with this startling information, President Roosevelt, in Washington, had asked Congress for the mightiest defense budget ever voted—some two billion dollars. As it turned out, the cost of making the first atomic bombs equalled the entire 1939 U.S. defense budget—two billion dollars.

The site of Klaproth's apothecary shop, now in East Berlin, where uranium was first discovered in 1789

Ernest Rutherford (right) with Hans Geiger—© *Schuster Laboratory, University of Manchester*

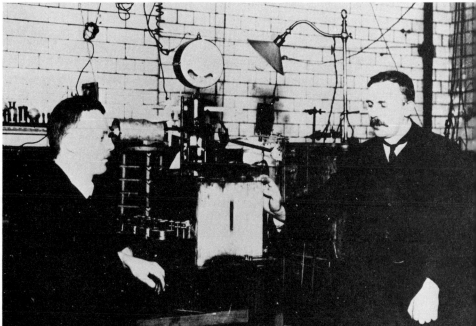

Marie and Pierre Curie in their laboratory—© *The Mansell Collection*

Lise Meitner and Otto Hahn early in their long association—
© *O. R. Frisch Collection*

The most powerful rock ever found—seven tons of the richest uranium in one lump, discovered at Shinkolobwe by Belgian mining engineers in 1922. It contained enough fissile material to build two Hiroshima-type bombs—© *Union Minière*

Left: Niels Bohr, the Danish genius who broke the news of uranium fission to American scientists—© *Lotte Meitner*

Below: The French team that discovered chain reaction in Paris, March 1939. Clockwise from the right: Hans von Halban, Francis Perrin, Frédéric Joliot-Curie, Lew Kowarski—*Photo: French C.E.A.*

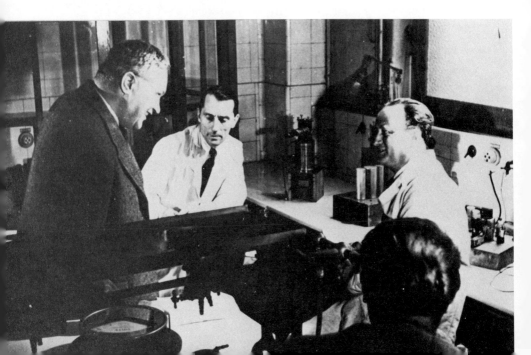

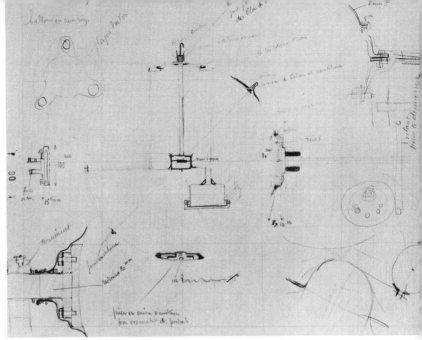

The first graph of a uranium reactor, drawn by Joliot-Curie in Paris in 1939—*Photo: Professor Lew Kowarski*

Vannevar Bush and Arthur Compton in 1940—*Photo: Lawrence Radiation Laboratory, Berkeley, California*

Los Alamos: Scientists listen intently to data on work of building atomic bombs: among them are Otto Frisch (front left), Enrico Fermi (middle front), and Robert Oppenheimer (middle of picture)—© *Los Alamos Photo Laboratory*

Otto Frisch, Rudolph Peierls, and John Cockcroft (left to right), after award of the U.S. Medal of Freedom in 1946—© *P. A. Reuter*

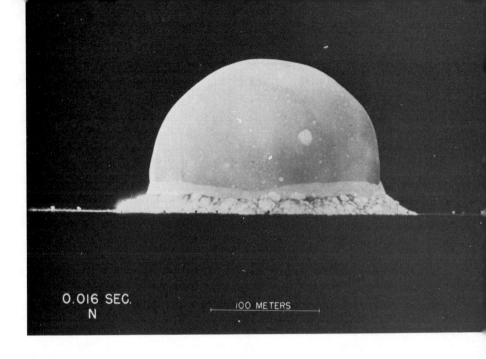

0.016 SEC.
N

100 METERS

The first atomic explosion, 16-thousandths and 34-thousandths of a second after detonation—*Photos: Julian E. Mack, Los Alamos Scientific Laboratory*

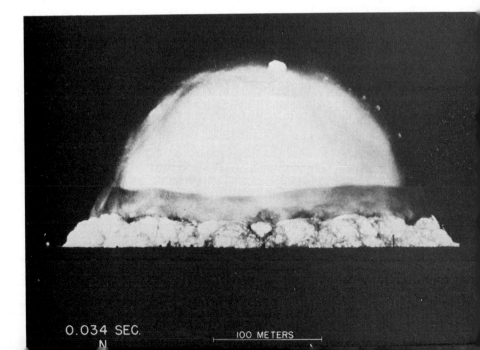

0.034 SEC.
N

100 METERS

Ernest O. Lawrence, Glenn Seaborg, and Robert Oppenheimer in early 1946 at the controls to the magnet of the 184-inch cyclotron.—*Photo: Lawrence Berkeley Laboratory, University of California*

Sir James Chadwick (discoverer of the neutron) with Sir John Cockcroft, at a reunion long after World War II—© *O. R. Frisch Collection*

THIRD PHASE
Two—Three—Five

On Thursday, January 26, 1938, some 12 hours before Niels Bohr threw the fission bombshell into the Washington conference, the city of Paris woke to an icy morning as Lew Kowarski hurried toward the Collège de France.

Kowarski was a bulky young Russian, a half-Jew, who had been hustled from Petrograd to escape Bolshevik repression in 1917. It had been a hard life, in Poland, Belgium, and Paris, working in factories and as a delivery boy; then he had served as an unpaid hand in the laboratory run by Francis Perrin and, from there, had graduated to personal assistant to Frédéric Joliot-Curie at the Collège de France.

Kowarski had a gift unusual in a scientist—he had a sense of history. It was this that led him through the snow to reach the laboratory much earlier than usual; he felt certain he was to witness an historic experiment.

Joliot-Curie had left his home in Parc de Sceaux soon after dawn; he was already at the laboratory when Kowarski arrived, putting finishing touches to his apparatus; he smiled a quiet greeting.

It was a physical test he had planned; a brilliantly simple device: a four-centimeter piece of brass tubing, coated with uranium paint; a similar cut of tubing without the uranium; several bakelite cylinders, shaped like simple napkin rings, each slightly larger in diameter, and the smallest a few millimeters larger than the brass tubing.

Kowarski sat, watched, and waited.

Joliot-Curie extracted a small neutron source from a leaden pot—the usual mixture of beryllium and radium—and placed it inside the unpainted brass tube; around this he set the smallest bakelite ring. After a brief while he lifted the outer ring and placed it near a Geiger counter. There was no reaction. He replaced the bakelite ring and removed the brass tubing, leaving the bakelite surface exposed to the neutron source; still no reaction. Now, he placed the neutron source inside the *uranium-coated* brass tubing and set the bakelite ring around this. He let a few minutes pass; then he removed the bakelite ring and placed it near the counter.

The Geiger counter clicked furiously. Fragments of shattered uranium atoms adhering to the inner side of the bakelite ring threw off bursts of radiant energy—pulsing proof of atomic disintegration.

It was clear that neutrons from the source inside the tube caused the uranium atoms to rupture so that new atoms of middle-order elements were created. Now Joliot-Curie had to establish the range of their radiant energy; this, too, was simply done. He kept changing the size of the bakelite rings, increasing the distance between the uranium coat on the brass tube and the collecting face of the rings until, at a distance of three centimeters, he could detect no more radioactivity.

"Bravo, bravo, Joliot! The neutron rips the uranium atom apart. A wonderful discovery!"

The Frenchman smiled, and said quietly, "Let's not get too excited. The Berlin report is there for all to see; any competent physicist could already have done a similar experiment. Maybe we are not the first."

Joliot-Curie was right; he was two weeks behind Frisch and an hour or two behind a test in New York, performed at Columbia University.

As soon as his report was written, for the French Academy of Sciences and the London magazine *Nature,* he called his colleagues into conference.

This new knowledge, of uranium disintegration induced by neutrons, he told them, was *not* the critical issue. The essential criterion to liberation of energy from uranium on a large scale was whether, in splitting the atom, other neutrons were released. If that was so, then, knowing *how many new neutrons* emerged from the shattered uranium atom would be vital to any hope of atomic power. He said, "We must try to show whether uranium chain reaction is possible."

The work would start at once; he would recall from holiday in Switzerland their colleague who knew most about neutrons, who had worked with Bohr and Frisch in Copenhagen, Dr. Hans von Halban. Francis Perrin would be theoretical consultant. The experimental team in the project would be Joliot-Curie, Kowarski, and von Halban, with Irène as consultant.

At the very hour the French group committed themselves to research chain reaction in uranium, the German foreign minister, Joachim von Ribbentrop, left Warsaw after attending the fifth anniversary of the German-Polish Non-Aggression Pact. He had found a stiffening of Polish resistance to German demands that Poland should join the Axis. In the train to Berlin, Ribbentrop told his aides, "We now have only one choice to escape encirclement. We must get an understanding with the Russians."

It was a setback to Nazi power diplomacy. Hitler had invented "cold war" and had given the Third Reich arrogant confidence from three great coups: the Rhineland had been taken; with *Anschluss,* Austria had been made part of Greater Germany; and, after the Munich sellout by Britain and France, the Sudetenland had been seized. Poland and France were now using "cold war" tactics, but soon the whole of Czechoslovakia would be seized, with its modern arms factories, its steel works, refineries, and its raw materials—including the uranium mines at Joachimsthal. Thus, the Paris team worked against a backdrop of threatening aggression. Suave and self-assured, Hans von Halban, son of a German professor of chemistry, returned from the Swiss ski slopes to the Collège de France to find work already started on the counting of neutrons.

His ideas at once clashed with those of Kowarski. "We can count the neutrons going in, and we can count them going out. Then, we need only calculate the difference."

"I think secondary neutrons should show a different character to the primary ones," Kowarski offered. "I think we should seek the difference in quality." He saw the stiffening of von Halban's jaw and joked, "Of course, Hans, you were taught neutrons by Frisch, and you taught me. So I suppose that makes Frisch my neutron grandfather. But I still think we should tackle this problem by quality, not quantity."

Joliot-Curie intervened and said they would do it both ways. There seemed plenty of time, then; they had a false sense of being alone in a new leap in science.

Then the February 11th issue of *Nature* came into the Collège, bringing the Frisch-Meitner report, and they heard for the first time of the term *fission*. Now they knew they were caught in a race for distinction, that the competition was international and fierce.

They raced on with their work, and then came a further bombshell, from New York—a letter from Dr. Leo Szilard, dated February 2: "About two weeks ago, a few of us here got interested in the possible release of neutrons from uranium disintegration ... if more than one neutron is liberated ... this might lead to the construction of bombs."

He went on to say it might damage mankind to make such data known: "I hope with all my heart there will not be enough neutrons released. But, if there *are,* we must not print results that would put awesome new weapons in the hands of the war-makers .... Please don't publish!"

Joliot-Curie discussed the letter with his group. They were being asked, he said, to obliterate their names in the history of science. How true was it that this might do harm? Halban was scathingly skeptical, aggressively Teutonic: "They are a bunch of scared conspirators! Sit still? When everyone over there is racing into print? We would be mad!"

Kowarski declared, "Someone will be first; why not us? It's childish to think the great German state or OGPU won't get this information anyway. They may call us names—but we shall be in honored company. Let's go into print."

They had not so intended, but now they decided to rush a preliminary report to London. They used their cloud chambers, ionization detectors, counters; they worked twelve-hour stretches. Early in March, they reviewed their data; they were then certain that secondary neutrons, stored in the uranium nucleus for all of time, were liberated with fission.

The final answer to Joliot-Curie's question—how many?—was not at hand; but they decided not to wait. They had a prize for neglected French science; they had the Curie name to uphold. They wrote their brief report.

Joliot-Curie hired a Peugeot, had it standing by the door with engine ticking over while the last lines were typed. Kowarski, still licking the envelope, climbed into the car to race to Le Bourget airfield. Joliot-Curie's last instruction was, "See it into the London bag, Lew. Don't trust any minor official. See the plane take off before you leave."

The report carried the heading "Liberation of Neutrons in Explosions of Uranium."

It reached the office of *Nature* on March 8; it was published ten days later. It said, in part, that release of more than one neutron was necessary for a chain reaction to be possible in uranium, and stated, "This seems to be the case. A chain of energy *is* possible."

As the London presses were printing this report, the German army moved into Czechoslovakia proper, to commit what has since become known as the "Rape of Prague."

When spring of 1939 came to Paris, the Collège de France team had achieved their fission-neutron equation. In the haste to get their results into print, error crept into calculations. However, this did not change the fact that their paper was the first with numerical proof that uranium fission released more neutrons. Chain reaction was no longer a dream.

Kowarski again served as courier to rush the Paris letter to the London-bound plane for inclusion in the April issue of *Nature*. Years after, he told this writer, "That paper had historical impact. It set alarm bells ringing in the major capitals. It roused instant action in Britain. It was read avidly by Kapitsa in Moscow. It had its effect in Paris—but, it did nothing in Washington. There, Enrico Fermi had already tried to alert the U.S. Navy but had aroused no interest, no reaction, and no immediate progress was made."

In Paris, however, the potential in uranium was a floodtide that caught them up in constant work. They pushed their investigation into areas of controlling atomic energy, into materials for slowing down neutrons; they put French atomic science once again—though briefly—in the forefront of the world.

Kowarski recalled, "I consider myself one of the discoverers of uranium chain reaction, but for such a status the philosophical approval of other scientists is needed. On that same basis I made repeated pronouncements that I did not go along with British and American circles that credited Hahn and Strassmann as discoverers of uranium fission. *They most certainly were not.* The West, at that time, had fixed ideas that it was German chemistry that led the world; yet the Berlin people did no more than extend what Irène and Savitch had reported the previous year. I am convinced that the true discoverers of fission were Otto Frisch and his Austrian aunt; and their work was as beautifully convincing as was Chadwick's experiment in proving the existence of the neutron and as the discovery of chain reaction in uranium was with us!"

In Paris, not knowing what Szilard had done in London in 1934, they applied for patents which they thought to be the first ever sought on atomic energy processes. They prepared for the Academy of Sciences a statement of their discoveries, and possible applications, which foreshadowed developments in the production of nuclear power; but by the time this was done, war was on the doorstep of France, and the document was hidden until peace returned.

· 2 ·

By the time exploding uranium atoms were revealing the promise of chain reaction to Joliot-Curie and his team, the Shinkolobwe mine had yielded a thousand grams of radium.

Ironically, when chain reaction was discovered in Paris, the mine was closed down. It was getting hard to sell radium, even though the Great Bear Lake deposit discovery in Arctic Canada produced competition which lowered the price by two-thirds to around $25,000 a gram. The market was saturated; outlets, in hospitals, research centers, with clock-makers and the "health food" entrepreneurs, were closing. The Belgians held a stock of 150 grams, a holding then worth about $2 million. This radium was stored in concrete bunkers the company had built, living out its long half-life of 1,600 years.

In their Brussels headquarters, board members of the Union Minière du Haut Katanga mining company had agreed with the president, Edgar Sengier, it was time to close Shinkolobwe until the market improved. More than 2,000 tons of hand-picked rich ore waited on the surface in the Congo. The Sharp reef had been worked below ground level and so would fill with rainwater. But when the time came, it would be easy to renew extraction; meanwhile, the native workers could be employed at other company mines.

There was no immediate market for radium and uranium ore was itself an encumbrance; Sengier was a far-seeing tycoon and a trained geologist, and his decision to shut down the operations at Shinkolobwe was made before Joliot-Curie's experiment.

It was not long, however, before a new view took shape in Sengier's mind. From Paris, Frederick Joliot-Curie, of previous acquaintance, called him early in April, 1939, to describe the new potential in uranium, to ask for assurances that supplies of refined ore would be available for his future work. Joliot-Curie also said he would come to Brussels soon to discuss possible joint action for development of atomic energy.

· 3 ·

In London through the first three weeks of April, 1939, Professor Sir George Thomson had watched spring emerge below his window in Imperial College.

In the parks, he knew, men were digging trenches while others were piling sandbags around buildings against the threat of air attack. The loom of impending war had marred the promise of summer—and now across his desk had come the shadow of the atomic bomb.

Thomson pored carefully over the figures quoted in the April 22 issue of *Nature*. The facts carried a chill foreboding. The French workers had calculated that when a uranium atom exploded from the intake of a neutron, *an average of 3.5 fresh neutrons were liberated.* Nothing stood in the way of chain reaction! There were other data—on slowing down neutrons, on the size of uranium mass in which this might be done—but there was the implied threat: chain reaction, with vast unleashing of energy. And it could be read at this moment in Berlin.

Thomson was the son of Sir J. J. Thomson, Rutherford's teacher and discoverer of the electron. He knew the chances; he knew there were limited amounts of radium-bearing ores in the world, that the largest holdings were in the hands of the Belgians. It was natural, in his concern, that he should seek advice from the man who had succeeded Rutherford at the Cavendish, Lawrence Bragg.

There was no hesitation in Bragg's mind; he saw good grounds for Thomson's anxiety. He was a brilliant physicist, the youngest ever Nobel laureate. He at once suggested consulting

Sir Will Spens, Master of Corpus Christi College; and Spens—
who was a civil defense chief—alerted Kenneth Pickthorn, MP
for Cambridge University.

It was Friday, April 22, 1939, when Pickthorn took a taxi
from Westminster to see Thomson at Imperial College, Ken-
sington. It was the day he was to prod Britain into the first
unwilling steps toward development of nuclear energy.

Pickthorn brooded over Thomson's warning during the week-
end, and on the Monday, talked with Major General H. L. Ismay
(later, Lord Ismay) who was Secretary to the Committee of
Imperial Defence. Ismay, no less amazed than Pickthorn, asked
the MP to dictate a note on what he had learned from Thomson.

Pickthorn's note opened the British government's "Most Secret" dossier on uranium. He stated that both Bragg and Thomson rated it "an even chance" that chain reaction could be
achieved in the element with a mass of sufficient size, that it
would be "a source of heat and power inconceivably greater than
anything yet known." The heat given off would be "100 million
times greater than that of coal . . . that it might have an explosive
quality greater—very much greater—than any known explosive."

Pickthorn also made a point that set in motion official bun-
gling, guile and hesitancy, which gave Hitler a chance to become
atomic master of the world. He reported that "somewhere in
Belgium is a great deal of uranium—from which radium has been
extracted—lying about, more or less a waste product."

German scientists would be alive to this prospect, the MP
warned, and so it was "a question of great urgency to know
where the uranium was, how much there was—and whether it
ought not to be got further away from a possible German grasp,
and into our own." It was, Pickthorn declared, a situation need-
ing serious attention, "a matter of days, if not hours."

It was far too much for Ismay's soldierly mind to deal with,
alone. Faced with an aura of scientific mysticism, urgency, and
invisible danger, General Ismay turned to the respected scientific
oracle among government advisers, to Sir Henry Tizard, chairman of the Committee on the Scientific Survey of Air Defence.
Tizard was a man of vision and, as one of the first official
scientific entrepreneurs, had promoted the secret work on radar,
a British invention.

Now, unfortunately, Tizard's vaunted vision sadly lapsed;
against Thomson and Bragg's "even chance" for liberation of

atomic energy, Tizard rated the odds at a long 100,000 to one! He cast the first cloud of doubt, and this had effect on all that followed. Still, he brought in the Foreign Office, the Ministry of Defence, and then the Treasury, for money that might be needed.

By April 26, almost a week since Thomson had read the Paris report, a senior section head of the Foreign Office, R. C. Skrine Stevenson, despatched a "coded cypher" to the British ambassador in Brussels, Sir R. Clive; it pointed out the possible value of uranium as a source of energy, asked that discreet soundings be made on available ore stocks, the chance of an option, the approximate cost, and added, "Commercial Counsellor should let it be understood that he is making enquiries on behalf of trade interests in the United Kingdom. Matter is urgent as it is thought possible that Germany may be alive to the potential value of this ore."

The possibility of the president of Union Minière also being alive to uranium's emerging value did not seem to occur to the British officials.

Mr. Pickthorn had other views. He called on Ismay that same day and declared again that Britain could "not afford to waste even hours on preliminary reconnaissance." And Ismay told his minister, Lord Chatsfield, that there should be a direct and open approach to the Belgian government—to seek the extent of Belgian uranium stocks, for "permission to ship the stuff to England at once"—and further suggested that "if the slender hopes were fulfilled, there should be a gentleman's agreement on financial compensation."

It did no good. By April 29, Skrine Stevenson had his first reply from Brussels. Clive reported Union Minière were sole principals; that some ores, about 65 tons, were with agents in London, and the price, ex-refinery, was six shillings and four pence per pound. The report added, "Information came through the President of the British Chamber of Commerce in Brussels, who represented the enquiry as originating from the Chamber of Commerce in U.K. . . . On this basis, an agreement of Union Minière was obtained for an option to purchase the entire stock in Belgium—provided definite assurances were forthcoming that uranium is required for national use, and not for purchase by merchants or speculators. . . ."

Messages from Brussels also revealed that an English peer,

Lord Stonehaven, was a director and vice-president of Union Minière. This news at once aroused speculation as to how Stonehaven could be used to ensure control of the Belgian uranium—without it costing too much.

Another week went by; then a meeting was called of Ismay, Tizard, Skrine Stevenson, and the Treasury man, B. W. Gilbert. They gathered in the afternoon at Ismay's office on Richmond Terrace; and Tizard at once threw a wet blanket over the unseemly urgency that had been injected by Thomson, Bragg, and Pickthorn. The minutes of that secret meeting recorded his opinion that "... the immediate possibilities of uranium had been considerably exaggerated, that it was not justifiable to advise H.M. Government to incur very large expenditure on the purchase of all available stocks at the present time." Tizard proposed, and it was agreed, that a *single* ton of uranium ore should be bought "to enable experiments to go forward. And in view of the possible, though not probable, outcome ... the government should endeavour to obtain an option ... provided the terms were reasonable." No further action by the Foreign Office was required; but the approach was made to Lord Stonehaven.

Next day, on May 9, Ismay wrote to Tizard, "Uranium again! From Lord Stonehaven ..." The president of Union Minière would be in London the following day, and some "big gun should meet this exalted personage." Ismay went on: "This business is out of my boat. ... I suppose I couldn't persuade you to add to your many kindnesses by 'bearding the lion' yourself?"

Tizard was quite ready to do that; and his casual approach to this critical meeting was astonishing. He wrote to Ismay after the meeting, saying he had "met the President of Union Minière, whose name I did not catch." The Belgian tycoon had been helpful and pleasant—it seemed such a friendly chat. "I was quite frank. ... I did not see how I could be otherwise, if we wanted his help, and that was quite a good thing because it turned out, in conversation, that he had had precisely the same information from friends in France. He is in close personal touch with the Joliot-Curies, who are investigating the uranium problem on exactly the same lines as is being done in this country."

Tizard, apparently, provided an escape hatch for Sengier; he did not ask for *uranium ores:* he asked for "black uranium oxide." Sengier replied there was no such material available in Belgium; that if Britain wanted to buy large quantities, Union Minière would have to enlarge its plant, but he could give no estimate of cost for this.

Sengier went further: he told Tizard he was willing to help, but could not reserve stocks for Britain since his friends in France and Belgium would need a supply. He also revealed he had shipped 200 tons of ores to America to meet normal commercial demands for the next two or three years and not because he heard the rumble of the drums of war.

In later years Sengier recalled that Tizard took him by the arm and said to him quietly, and seriously, "What we are most concerned with is that no country with which we might be at war in the near future should be able to corner all existing stocks of uranium mineral ores."

Tizard recorded that Sengier suggested it might be desirable to move uranium stocks from Belgium to Britain: "I said I had no authority, but thought it highly probable the government would be prepared to store such material, if he wished. It would not take up much room."

Tizard mentioned a need for a ton or two of oxide and closed his note to Ismay with the opinion that "I rather doubt it worth-while to get a rigid agreement on paper. . . . If the government tries to press him now on questions of supply and costs, he may get restive."

And, Tizard added, he still thought the supply of uranium to Britain should be "a kind of gentleman's understanding."

Tizard was content to see Sengier depart with nothing defined, with nothing arranged on removing the uranium from its site so close to the German border to a place of safety. He left it as a "gentleman's understanding"—with a man whose name he did not catch.

Tizard's words were taken by Sengier as a warning; they left him more than ever convinced of the possible role that awaited the waste dumps at Oolen and the still radium-rich stockpile of ore at Shinkolobwe. His discussions with Joliot-Curie had been more explicit and explanatory than he had been prepared to divulge. Sengier had already planned a meeting in Brussels with Joliot-Curie and some of his closest colleagues. They were to explore ways to set up a center in the safety of the Sahara to develop a joint French-Belgian effort to harness the power in uranium, and, perhaps, to use that power as a weapon of war.

All through the years of radium production, the Union Minière plant, at Oolen, had maintained the three tons-one gram ratio; this meant that, for every gram of radium produced, three tons of uranium concentrates had been thrown on the waste dump. Later in that summer of 1939, it would evolve, from studies in Britain and America, that in the piles of yellowing waste at Oolen was enough power to devastate dozens of major cities.

The date of Tizard's meeting with Sengier in London was May 10, 1939. Exactly a year later, with the yellow piles still standing behind the factory at Oolen, Germany invaded Belgium. Soon, scientific officers of a special ordnance division set up by the Wehrmacht were at Oolen to confiscate more than 1,000 tons of the ores from Shinkolobwe, a supply equal to that which was to save the American atomic bomb project from being moribund.

· 4 ·

Niels Bohr returned to Copenhagen from America even as the letters were flowing between London and Brussels. In those few spring days of the Atlantic crossing, Bohr was exploring a new idea. At Princeton University, once the flurry of excitement had simmered down over the fission disclosure, Bohr had detected in various experiments clues that intrigued him. He was always curious about coincidences in nature's symmetry and laws, and there was a coincidence here. It aroused a suspicion that the beautiful simplicity created in the liquid-drop uranium nucleus, by the arrival of an extra neutron, was misleading; he felt there was *a something* they had all missed. He sat down with John Wheeler to work out the theory.

Bohr told Wheeler, "It is obvious that neutrons cause fission—but not in *all* uranium atoms. If that did happen, then all our uranium target atoms would fission, and we should have a much greater release of energy . . . so, exactly how much disintegrates? The figure indicates something a little below one per cent—and that will tell us to suspect it is *not* the ordinary atom which fissions at all. It is the uranium *isotope.*"

Unleashing the atom's energies was, from now on, a mere

elucidation, a technological formality to attain the stage Ernest Rutherford had foreseen in his speech in 1916—release of nuclear energy at will.

Through the last weeks of winter, 1939, Bohr and Wheeler worked on the coincidence theory. The more they sought its falsehood, the more it looked to be right. In those weeks, after years of being rated by Otto Hahn as "too rare to matter," Arthur Dempster's 235 isotope was, suddenly, a central character in the atomic drama. A quarter of a century later the consequence of Bohr and Wheeler's role was noted when, during the presentation of a national award to Wheeler, a Presidential citation described how the paper which he wrote jointly with Bohr had been "the cornerstone for all later understanding in this field, its publication a step forward to unlocking the fantastic secrets of the nuclear age. . . ."

The liquid-drop nucleus was the vital clue. The compact mass of 92 protons—as there was in uranium-238—jostling with 143 neutrons in that incredibly tiny space, suffered a critical instability with the intake of a slow neutron—an instability that did not arise when an extra neutron entered a 238 atom, with its 146 neutrons! The resonance, the quiver of the droplet core, the breaking apart into two atoms of middle-order elements, happened only in the "rare" 235 nucleus of uranium. Seven atoms only, in a thousand of the stable 238! It presented an astonishing picture to the science of physics, still reeling from the fission disclosure.

On his way home to Denmark Bohr sketched the main form of his ideas for a report. He and Wheeler agreed on the contents by correspondence, and their paper was finally sent, in summer, to the editor of the American *Physical Review*. It was a classical analysis, titled "The Mechanism of Nuclear Fission." Despite the fears and the pleading of exiled European physicists in America for secrecy, it was published—24 hours before German tanks rolled into Poland to start World War II.

Incredibly, the notion of a neutron-induced chain reaction in uranium-235 at first escaped the attention of the main protagonists of the fission discovery. Nor did the possibility of a slow reaction in common uranium at first attract people like Hahn, Strassmann, Meitner, Frisch—and many others. They continued to sort all the various fission fragments from uranium disintegration into their proper places as isotopes—no longer as trans-uranium substances, or offshoots of radium. This was pure science in action.

However, one man in Berlin, Dr. Siegfried Flügge, a colleague of Hahn's in the Kaiser Wilhelm Institute for Chemistry, did take notice of the implications. He started work on theoretical ideas before Bohr's return journey to Denmark. Flügge did not have the 235 isotope data as a foundation. He started with the same assumption of the people in Paris that two or three neutrons would be released in fission of each uranium atom. He thought this liberation of particles could be made at will, and he reasoned a way might be found to liberate the energy in every single atom in a large block of the element. He calculated, "If you take a large volume of uranium oxide equal to a cubic meter, and if you achieve a total reaction with this flow of neutrons, there would be a release of enough energy to lift a cubic kilometer of water, weighing a billion tons, to a height, above the earth, of seventeen miles."

This astonishing sum was too great a jump for radiochemists used to dealing in millionths of a gram; it aroused no fever of excitement in Germany. The remarkable paper, written without knowledge of the critical role of the uranium isotope, was printed by the editor of *Naturwissenschaften* in June, 1939. Its title asked a question not to be answered in Hitler's Germany: "Can the Energy of Atomic Nuclei Be Used in Technology?"

The report had far greater effect in the West. There, essential data was already available from which scientists and engineers and technologists in America would achieve the explosive energy actually to lift a cubic kilometer of water above the earth. It added to fears, already spreading in Britain, France, and America, that physicists were working toward a nuclear-armed Germany.

The miasma clouding the leading minds in Berlin also existed in Copenhagen. Lise Meitner came over from Stockholm, where she was working at the Nobel Institute, to collaborate with Frisch on fission by-products. Working together, they measured the recoil in atomic fragments caused by intake of neutrons. Long after, Frisch noted, "And, in the excitement of this new work, we missed the critical factor—the chain reaction."

This did not arise in his mind until, in discussion, the Danish theoretician, Christian Moeller, said, "I've been working on energy levels of fission of uranium. It looks as though there are secondary neutrons, Otto! There might be a possible chain reaction."

The Paris work was not yet in print; Bohr was still to leave Princeton with his ideas of uranium-235; and Frisch's first reaction was incredulity. He protested, "How could that ever be? If that could happen, all the uranium deposits in the world would long ago have blown up! "

Even if the neutrons did multiply, he reasoned, they would be soon swallowed up in the other elements mixed in with the natural uranium. But wait! Isn't that just *why* these explosions do *not* occur in natural deposits. Yes! The argument against chain reaction on those grounds was too naive.

Gather together enough *pure* uranium and, with care, there could be a way to achieve a controlled chain reaction and liberation of nuclear energy on a scale that mattered, that could be useful. Yet there was a shadow with this reasoning, a specter that a continued chain reaction could extend to explosives in war. But it was still compelling.

On Bohr's return to the Institute, apologies for premature disclosure of the fission secret, Rosenfeld's explanation, were dismissed by Frisch as "imprudent communication." Now there was this immensely challenging concept of the isotope being the key factor in the fission process, and again Frisch found a new idea implausible—at first; he noted, "Bohr's argument was by no means conclusive. It did not convince everyone."

The rare isotope would have to be separated out before experimental evidence could settle the matter; and there lay his challenge. It came at a most disturbing time for Frisch. He said, "I did not feel like starting any new important work under conditions of renewed threat. Denmark was in danger of invasion. I felt it would be wise to move again. I went to England in the summer, and I was there when war broke out."

From the Union Minière headquarters in the Montagne du Parc, in Brussels, Edgar Sengier compounded the Tizard mistake of seeking "black oxide of uranium." Six days after the London meeting, he wrote Tizard that, as the demand was very small, the stock in Belgium was no more than "a few tons divided in different places." He went on, "We could start deliveries of between 15 to 20 tons a month on a firm order, but not before next November." The company could then only supply about 100 tons in the first six months, and the price would be close to ten shillings a pound.

There was no offer of a total option; there was no mention of the thousands of tons of refined ore at Oolen. There was an inference Britain could not hope for priority: "We have a number of contacts we must continue to supply first; also, for years we have been in close cooperation with scientists of the Curie family who, as you know, have been carrying on experiments, referred to in our interview, with products we have supplied."

When this letter reached London an even colder wind blew over the British uranium prospect. It came from rare heights, from the chief of the Department of Scientific Research and Experiment at the Admiralty—that same august sector of defense to which Leo Szilard, in 1934, had donated his chain reaction patent to preserve it from misuse. C. S. Wright kept faith with that altruistic gesture; he poured instant cold water on the idea and disparaged reports running wild in the British press that Hitler's "secret weapons" included a superbomb of a few pounds of uranium that could destroy London.

Wright expressed his sarcasm succinctly. He wrote to General Ismay on proposals to buy up the Belgian uranium, "My first reaction . . . is to suspect that someone has something to sell; and it is extraordinary how well-meaning people (e.g. Mr. Pickthorn, who knows nothing of uranium) unconsciously further schemes of this kind." He added to his suspicion of Union Minière his conclusion that the reason why other countries were not rushing to buy the uranium was because "the possibility of developing an explosive of unprecedented power from uranium is so remote as to be negligible . . . and we should be well advised to hold off any large expenditure to buy up stocks of uranium waste products."

In contrast to the Admiralty's attempt to torpedo the British uranium project, work was already under way on the Paris project. Sengier was arranging a supply of 70 tons of refined ores, in nine railroad cars, to back the French plan for experimental work in the Collège de France and at a yet-to-be-selected site in the

Sahara. At the same time in London Sir George Thomson had acquired two tons of ore for tests while the professor of physics in Birmingham, Mark Oliphant, also had a ton of uranium which he would "use to find the best way to make metal—in the spare time his work on radar allowed."

Then out of Berlin came a story of a national conference of nuclear physicists reportedly discussing the uranium problem. Word of this reached the professor of metallurgy at Cambridge, Dr. Hutton—from a "reliable source in Germany"—that experiments were planned with 100 kilograms of uranium and that Nazi authorities were "taking very seriously" the question of an explosion.

Sir Henry Tizard was informed of all this. It did nothing to change his view. After a lapse of three weeks, he replied to Sengier's letter saying this delay was a "mistake"; and he put off any question of uranium supply to Britain until experimental work was done in Birmingahm and London. By now, Tizard had all of the British uranium eggs in his ministry basket on grounds of preventing overlapping with other departments.

By the end of June, 1939, with war two months away, his opposition to the prospect of atomic explosives had hardened. A Cabinet note in the "Most Secret" files read, "Generally speaking, Sir Henry Tizard said further investigations of this matter, made by the Air Ministry Committee, had confirmed his view that its importance had been exaggerated."

All the same, he still had an each-way bet: he agreed to continue experiments in the hope the scientific picture would be clearer in two or three months; meanwhile, he thought the position "quite satisfactory."

Those next two months brought the opening of World War II; and instead of giving greater urgency to the uranium question, national mobilization brought about a hiatus.

In Paris, the Joliot-Curie team had neither reservations nor hesitation. In pragmatic fashion, they tackled problems of experimentation that stood between atomic energy as an abstruse idea and as a power source they could turn on and off at will. The term "atomic energy"—then strange-sounding to non-scientific ears—became a cliché in the Collège de France.

A week before Tizard in London had his meeting with the

head of Union Minière, the Paris group had already lodged the first two *Brevets d'Invention* with the French patents office. These included statements on systems for slowing down neutrons, for control of chain reaction in a nuclear reactor, and for critical parameters in assembling volumes of uranium from which energy could be extracted. Francis Perrin had produced the theory of the explosive critical mass. He stated there was a volume of uranium that had to be reached to prevent escape of the neutrons to the outside, which would cause the nuclear fire to fizzle out.

The agreement between the four workers, Joliot-Curie, Kowarski, von Halban, and Perrin, to seek patents was alien to the dictum established by Pierre and Marie Curie. Madame had written on how, with the discovery of radium and polonium, she and Pierre had decided to publish without reserve and had "renounced any attempt to reap profit from our discovery." The decision to apply for patents in 1939 came after much soul-searching and only on the grounds that patent rights would give the scientists a voice in the great issues the new power source was likely to bring. They had put French nuclear science in a position of world leadership by their vision and their industry. Kowarski afterward recalled, "In the weeks prior to the declaration of war, we worked twelve-hour days, building essential understanding of the process and nature of fission phenomena. We had to investigate moderators for slowing neutrons, and we showed that ordinary water would not work. We had to work our way through other moderators and develop theories on all the aspects we had in view. And this kept us very busy ... then, suddenly, war was on us, and Joliot-Curie was called up!"

In one stroke French atomic study was made a national undertaking. The head of the team was commissioned—Captain of the Artillery Frédéric Joliot-Curie; and all the laboratories came under military law. Captain Joliot-Curie followed orders—as far as they went. Secrecy fell over the work, laboratories were code-named. Kowarski typed out the changes; this room, previously the neutron study room, would now be Laboratory 2A, and this room Laboratory 3B. The "phoney war" was on, and the first way in which they were shackled was on priority for experimental materials; but even this did not slow their work. By October, 1939, they were arguing whether to go for graphite as the ideal moderator for the slow neutrons, which they saw were vital to chain-fission, or, whether the very rare isotope of hydrogen—the two-part atom that Rutherford had predicted—could be used.

They used the analogy of a type of billiard-ball collision between neutrons and hydrogen atoms that would not absorb neutrons. However, if the hydrogen was two-part, deuterium, then the billiard ball would be twice as heavy. The slowing effect on the neutron would be doubled! And there was a source of deuterium in "heavy water."

Heavy water! Three years later, in a meeting with President Roosevelt, Winston Churchill saw heavy water as a "sinister, eerie, unnatural term" that was then occurring in his secret papers. To deny its use to German physicists, commandos were parachuted into Norway.

As the first winter of war drew on, the Paris team worked toward basic fundamentals of the first uranium reactor. They were now keenly aware of the 235-isotope situation, and they looked carefully at the question of atomic explosives. They wrote a secret pronouncement that stated many of the facets that were to be closely guarded secrets in wartime America and Britain; they sealed the document and lodged it with the Academy of Sciences in Paris. Soon after, it went into a hidden vault of that society and remained there for five years while the Germans occupied the city.

• 5 •

Otto Frisch was technically an enemy alien. In Birmingham University he was barred, on the grounds of security, from entering physics laboratories, from knowing of the radar work being done therein. Yet he was freely permitted to work on the release of atomic energy. In this Gilbertian situation, he reached a conclusion that accurately forecast the way to the uranium bomb.

Frisch had been in Birmingham in September as Oliphant's guest, discussing his future when on a sunny Sunday morning the reedy voice of Neville Chamberlain sent Britain to war with Germany. Frisch said, "Oliphant extended the invitation and I just stayed on."

He was not a lone exile; there, also, was Professor Rudolph Peierls (pronounced Piles), a German-born physicist who had been abroad when Hitler came to power, who refused to return to Germany, and who took a chair in England. Peierls welcomed him to his home.

Before this, the idea of a super-bomb had been bandied about publicly, even though big guns in administration and in politics had shot down the idea. Winston Churchill had been among them. Advised by Lindemann (whom he afterwards created Lord Cherwell), he wrote to the government there was little chance of such a bomb, that its power could not be much greater than normal TNT, weight for weight. Skepticism was deep and wide but not profound, and uranium research was kept alive as a very slim prospect as the ranks of British scientists were bent to what Tizard saw as "more immediate research."

Three months after war broke out, ten months from the announcement of fission, Whitehall turned for the first time to the most eminent nuclear brain in the British scientific establishment, the man who had found the atomic bullet, the neutron—James Chadwick.

Chadwick was cautious; yet weighing the facts and stressing the lack of real data on the mechanism of fission, he wrote, "explosion is almost certain to occur if one had enough uranium." His estimate of "enough" ranged widely—from one ton to around 40 tons. He said he was sorry not to give a definite answer to this interesting question, and added, "The amount of energy which might be released is in the order of the well-known Siberian meteor. . . . The difficulty is lack of data. Very few experiments have been made; they have been concerned with radioproducts—apt to degenerate into a kind of botany. . . . I think if I can get enough uranium oxide to do some work on the fission mechanism, I will do so. . . ."

The doubt in official minds did not waver; the lords of the Admiralty were even reassured. Lord Hankey sent a note with Chadwick's assessment to the Admiral of the Fleet, Lord Chatfield, then Britain's Minister for Defence Co-ordination. It read, "Dear Ernie, I enclose a further letter . . . about the idea the Germans might use uranium against us. I gather that we may sleep fairly comfortably in our beds."

That was not to be. The alien scientists in the Midlands were soon to find means to realize uranium's stored energy.

As with many great events in history, the trigger to enlightenment was a simple matter. Otto Frisch was asked by the Chemical Society, in London, to write a report on uranium fission for their annual review. It caused him to re-examine the upheaval that had followed the first paper on uranium fission. He looked again at some of the results and prospects of enriching natural uranium with ten times the normal amount of uranium-235 so that, instead of seven atoms in every thousand there would be 70. And he, too, threw doubt on the chance of reaching a chain reaction with moderated neutrons; he used Bohr's current argument that this made efficient nuclear explosion impossible because it would give only a slow—thus ineffective—release of energy.

Frisch, then, was still not convinced by Bohr's reasonings that uranium-235 had a large ability to capture slow neutrons and was responsible for the observed effects. There was, he told himself, one way only to test the idea, and that was to separate a sample of the rare isotope from common uranium and strike it with neutrons. The device that first came to his mind to achieve separation of the isotope was a Clusius tube, a method of isolating, by a series of different temperatures, atoms of different weights. There was no chemical path he could follow; uranium-235 was chemically identical with its more numerous cousin because it had the same number of electrons, the same number of protons. Only by a marginally different response in gradients of heat would the lighter isotope show itself.

There was no Clusius tube available in Birmingham, and he was barred from the physics laboratories. In an unused lecture room he set to with a few materials and tools and started to build the instrument for himself. He pondered the method of having a gradient in temperature in the tube so that, when the uranium was turned to gas, the atoms would gather according to their weight, or mass. This was thermal diffusion; it could only be done with uranium made liquid at normal temperature and mixed with fluorine to form a stable but highly corrosive compound known as hexafluoride. And that would have to be obtained, or made.

As he pondered his article for the Chemical Society, it occurred to Frisch to question how many Clusius tubes would be needed to separate enough uranium-235 to make possible a chain reaction—and *not* with slow neutrons, as he had dealt with in his article, but with fast neutrons—with the new, high-energy neutrons, unleashed from fission action, striking at the surrounding uranium-235 atoms.

His memory now carried him back a year, to the Institute in Copenhagen, and the world's first fission experiment, and writing to his mother of his experiment: "I feel like a man in the jungle who has caught an elephant by the tail and does not know what to do with it."

But now he could turn to the theoretical brilliance of Rudolph Peierls, who had already expanded the critical mass idea which Francis Perrin had advanced. Peierls had a rough formula for a critical assembly of uranium and parameters for neutron capture in uranium-235 atoms. He and Frisch then made guesses; they assumed nuclear constants yet to be proved on the true number of neutrons released from each fission. Working late at night over several weeks in Peierls' house at Edgbaston, they came to the first picture of nuclear reality. Said Frisch, "We were absolutely staggered to get the result that a couple of pounds of uranium-235 would be enough to make an atomic bomb. If my Clusius tube worked, it would take only about 100,000 of them to produce enough separation of the isotope, and the cost would be minimal in the national budget for the war."

In the aftermath, the British nuclear historian, Professor Margaret Gowing, commented on their written conclusion: ". . . its grasp of principles and properties is a remarkable example of scientific breadth and insight. It stands as the first memorandum in any country which told of making a bomb and the horrors it would bring."

Dr. Otto Frisch had caught an even bigger nuclear elephant by the tail; and again it was time to consult someone of wider experience.

· 6 ·

Head down against a cold March wind, Captain Joliot-Curie strode across the Place Marcellin Berthelot, his thoughts fixed on a design for the world's first experimental nuclear reactor. The year since they knew chain reaction was feasible had been spent in the search for the true moderator, for the right material that would be the medium for slowing down fission neutrons to give the greatest chance of capture.

The emergency did not seem so acute as when, in September, the bureaucrats had run haywire, closing down the Métro stations, canceling transport, calling up millions of Frenchmen. Now it seemed a false peace.

In the weeks since Christmas, they had worked with their own sense of urgency. Sengier had told him of the British diversion and that another director of Union Minière was the Belgian ambassador in London, Baron Cartier, who also knew of Whitehall's interest in uranium.

In the basement of the Collège the French team had erected simple experimental reactors; they pressed blocks of the Belgian uranium and placed these among structures of graphite. The uranium blocks were 15 inches by 3 inches and were built into a sphere inside the graphite. It did not work. There was no measurable chain reaction. They tried dousing the whole assembly with water, normal water, to slow down the neutrons. Again, there was no reaction. So they came to the idea of using heavy water, with the uranium powder loose, diffused in the liquid.

Now Joliot-Curie entered his office and took from his desk a large empty envelope that had come from the National Center for Scientific Research, to which his laboratory was now attached for national service. He ripped it into two pieces to use as sketch paper. He made the first rough notes and drew the outline for a nuclear reactor assembly.

On one piece of the envelope he scribbled the essential features; on the other piece he drew the outline of the bearings. He wrote in a comment, "We have to disperse the oxide powder through the water; so the reactor will have to turn on a spindle. It will be a sphere, and it will rotate—like a balloon." An atomic balloon!

Attached to the revolving spindle would be a paddle to stir up the uranium powder to prevent it settling. There would be no controls. This would not be a sustaining chain reaction, so it would need no controls. It was to be used merely to show that energy could be won from uranium inside a vessel with the right moderator. Yes, he needed the right moderator—heavy water. Materials were hard to come by in this hangfire war, and there was not enough heavy water in all France, but there was enough elsewhere, in one place.

He wrote a memorandum to the minister for armament, Raoul Dautry, telling of his work and his need. He was called to the minister's office next day, was rebuked for having delayed so long, and asked for his requirements.

"This is a matter of national importance," said the minister. "We shall call in the secret service, and we will have to take special precautions."

The next day, four agents in plain dress called at the Collège and took Lew Kowarski and von Halban into custody. The two physicists were told, "You are foreign-born, and you will be kept under detention for your own good until certain events have taken place."

Kowarski was taken to a small island off the Brittany coast; von Halban was kept on an island off the south coast. Both were told they would be held in custody so that they could not be accused later by the Germans of complicity in the heavy water operation.

It was typical French pragmatism.

Jacques Allier held the post of loan officer in the Paris headquarters of the Banque de France. There his tall, lean figure, dressed usually in gray suit and quiet tie, was well known, and he was accepted for what he seemed to be.

In mid-March, 1940, Allier had a telephone call. He was asked to ring Jules. He did not do so; instead, he cleared his desk, put away his files and locked his room. Then he left the bank, called a taxi, and rode to the Quai d'Orsay. Now he resumed his true role—as an expert in explosives and a lieutenant in the Deuxième Bureau, the French secret service. On that same day, his chief took him to see the minister for armament. Dautry eyed him grimly and watchfully, then asked, "Do you know the term 'heavy water'?"

Of course, he knew. In his cover job he had been party to organizing an investment the Bank had in the hydroelectric development in Norway, near Rjukan, which had opened in 1934.

"It is the product of the Norwegian plant. They use their excess power to separate it from normal water by electrolysis, I understand, sir. It is the only industrial production of heavy water in the world. All else is tiny amounts made in various laboratories."

Dautry said, "Exactly. And what is it that now makes it so precious?" Allier said he did not know.

"And you are not even to ask, Lieutenant! You are to go immediately to Norway and obtain all the heavy water they have. But it is no longer to be called heavy water. You will know it only as Product Z. We don't want you to know why this is so precious for your nation because there are many Germans already in Norway. You could be captured and the truth extracted. It is a secret, important mission. It could be very dangerous. Product Z must not fall into German hands—at any cost . . . for the safety of France!"

Allier's chief asked him for a cover name to travel by; next morning, the passport was delivered to his home, and it was in the name of Kreiss, his mother's maiden name. Within 24 hours of his instructions from Dautry, M. Kreiss arrived in Oslo for a business trip, and Jacques Allier was plunged into the first espionage drama of the atomic age.

In his hotel room waited two agents from the French embassy; they said they had been placed under his command. One was Captain Mollier, the other, Lieutenant Mosse; he decided to call them Jean and Henri. He asked Jean to keep contact with the communications sector of the embassy; he told Henri to observe his (Allier's) movements at a distance to see whether he was being watched. Henri reported that he was obviously under observation. Jean brought news from the embassy that Paris had intercepted a message from Oslo to Berlin mentioning the name of Kreiss.

Evasive action was taken. In the very early hours of the morning, Jean brought an embassy car to the back door of the hotel; Henri was in another car. They sped to the western suburbs, to a point where Allier then waited in his car at the roadside for some minutes before driving slowly on to the highway leading west. After a while, Henri, in the car behind, flashed his headlights. With this go-ahead signal, Allier drove fast for the next 80 miles—to Rjukan, to the mountain site of the Norsk Hydro heavy water plant. He knew the road; he had made the journey several times, and he knew the general manager of the plant, Axel Aubert, well enough to rouse him from his bed.

They went to the factory, and Allier saw steam rising against the wooded slopes from the plant. Allier said he wanted all the heavy water trucked to Fornebu airfield, near Oslo. Aubert was intrigued: "The Germans are already bidding for it. I've had an approach from the I.G. Farben company. We have 165 liters in stock, stored in aluminium cans with cadmium lids to prevent unwanted reactions."

Allier asked for the cans to be packed into two wooden cases and for two other cases to be prepared, looking exactly the same, but filled with ordinary water; he then gave delivery instructions. Aubert said it would take a day or two to make the arrangements, and he would telephone the Embassy with a code message saying the stream had dried up, once the cases were in transit. He cautioned the Frenchman, "I tell you, M. Allier, we must not have money for this until the war is over. I know if France loses, and the Germans come, I will be shot."

Allier was back in Oslo by noon.

Through Jean, he made arrangements for the flight of the heavy water. Two days later, in the morning, Aubert telephoned the message, and Allier then went openly to the airline office and booked a seat, and two cases of luggage, to Paris.* He knew he was watched. When the call came, he boarded the aircraft, saw the two bogus wooden cases aboard, and took his seat. The plane taxied to the end of the runway; then the pilot opened the door, Allier swung out onto the runway where Henri waited with a car. They raced to the commercial airfield at Fornebu where a chartered DC3 stood waiting with Jean and the real heavy water cases aboard.

They took off at once and flew low to the west, passing over Rjukan on the way, straight out across the North Sea. The pilot took them into cloud cover, and the next land they saw was Scotland. They landed on a private strip at Montrose; a truck was waiting to take them to Edinburgh railway station. There the two cases were loaded into the baggage van, and they sat with them, revolvers under their armpits, all the way to London. Again they were met and taken to Croydon airport, south of London; from there a French military aircraft carried them and the heavy water to Paris. The British were not aware of what had passed through their country.

---

* Allier learned later this aircraft was forced by German fighters to land near Hamburg, and the two bogus cases of normal water were taken off.

Allier took the entire world stock of industrial heavy water straight to Joliot-Curie's laboratory. The scientist was thrilled. He threw his arms around Allier, kissed him on his cheeks, shook his hand.

"Allier, France owes you a big debt! You are my friend for life."

He led the agent into the basement laboratory. "Now you shall see why this water is so vital to our cause." He showed him a drawing of his plan for the heavy-water-moderated uranium test reactor.

"You see, it is to be a steel sphere, spun in two halves, locked together with an opening to insert the uranium and the water. I believe when we have built this, and operate it, we shall show how energy unlimited can be won from the element, uranium. It could be the way to a mighty bomb."

Allier looked grim; it was not all good news he brought, he said. The Germans had been after Norway's heavy water. It was obvious they, too, were working on atomic energy.

This intelligence did not surprise Joliot-Curie. What gave him more concern was news he had from the Nobel Institute, in Stockholm, a few days later. He took his worry, then, to Raoul Dautry.

Professor Niels Bohr reached Oslo on the final stages of a study tour of Scandinavia, during which he had spent some time in Stockholm. In Oslo, he dined at the royal palace with King Haakon. The glitter of this splendid setting did nothing to lighten the gloom. Bohr afterward told his son, Aage, "It was a royal dinner in a depressing atmosphere. The air was full of a sense of doom, and the talk was of impending attack. The King himself was certain that invasion by the Germans was imminent."

Bohr left for Copenhagen the next day, April 8.

He went to his home and called in a young English girl who had been serving his family as a governess for a year while she was studying. He told her of his fears and said she should leave the country at once; this she did.

The next day, the sky over the Danish capital throbbed with the roar of squadrons of Heinkels and the scream of Stuka dive-bombers. German tanks crossed the borders; fifth columnists helped the invaders seize the ports; the conquest of Norway began, also, that same day.

In Paris, as the sound of battle rolled across Denmark and Norway, Lieutenant Allier was again summoned to the office of Raoul Dautry. Joliot-Curie was there to give him a special briefing, and Dautry handed him a letter of introduction. He went straight to Le Bourget and claimed a seat in the first aircraft going to London.

It was a strange gathering—a French secret service agent and four stiff and rather formal British scientists. They met in a quiet room at the old Royal Society headquarters in Burlington House to hear the story of the flight of the heavy water. Facing Allier were Sir George Thomson—his latest work with uranium had lessened his hopes—Chadwick, who was still noncommittal, Dr. John Cockcroft, then with the Directorate of Scientific Research of the Ministry of Supply, and Dr. P. B. Moon, representing Oliphant.

Allier had disturbing news, other than his heavy water tale. He gave a review of the Paris research into atomic energy and then related the concern that had built up in Joliot-Curie's mind in recent days. It was an anxiety concerning Frisch.

As soon as the meeting was over, John Cockcroft went to the office of a departmental security chief, Wing Commander Elliot, and dictated a note that later passed into secret Cabinet documents.

Allier, Cockcroft reported, had come to discuss work being carried out by Joliot-Curie and his team in Paris on the "production of a uranium bomb." And, the note added, "Professor Joliot-Curie had heard that Dr. Otto Frisch was working under the direction of Professor Oliphant, in Birmingham, on the isotopes of uranium, with a similar object in mind. He understood Dr. Frisch had been in communication with various scientists on this question, and he was very perturbed about the possibility of leakage to Germany, particularly in that Dr. Frisch's aunt, Professor Meitner, is now working in Stockholm, and that she might have communication with her former collaborator, Hahn, in the Kaiser Wilhelm Institute, Berlin. He [Joliot-Curie] asked that steps should be taken to ensure secrecy. . . ."

Cockcroft also reported, "It is not at all clear that even with heavy water it will be possible to make a uranium bomb—but the chances are now appreciably increased, and Professor Joliot is working at high pressure on this question. . . ."

Allier's information virtually coincided with three typed fool-scap pages from the University of Birmingham. These had been sent, on the urging of Professor Mark Oliphant, to Sir Henry Tizard, who felt the content to be out of his depth and had passed them on to Sir George Thomson for comment. These pages were eventually to achieve fame as the "Frisch-Peierls Memorandum," which held out favorable prospects for separating the uranium isotopes and causing "an explosion of staggering proportions."

This remarkably accurate document combined with Allier's news to shatter the laissez-faire attitude of the comfortable British scientific and defense mandarins. Britain started to move into preeminence in uranium research, and yet, like France's, it was a position held only briefly, never to be regained.

After thinking about Cockcroft's note, Wing-Commander Elliot dictated a memo to his chief, General Ismay, recalling "the secret enquiries of last summer into the question of uranium."

He attached the note from Cockcroft on the Allier meeting and said he had arranged for Cockcroft to see Tizard that same afternoon. He added, "There remain, however, two further aspects which require consideration at once:

(i) . . . ensuring information does not leak to Germany through Professor Frisch . . . the Secret Service should be put on to this at once.

(ii) Belgium, you will remember, is the source of raw material—black oxide—for uranium, and although the stocks were said to be small last summer, it is perhaps worth ensuring at once that action is taken to make certain there is no possibility of these . . . falling into the hands of the Germans."

At this late juncture, Tizard joined the urgency. Almost a year after his meeting with Sengier, he told Lord Hankey, "It is essential that measures should be taken to ensure that everything which might be of use to the Germans in the event of their invading the Lowlands will be removed or destroyed."

It was far too late. The German buildup was nearly complete for the Mannstein plan, adopted by Hitler against High Command wishes. This was for a lightning strike through the Ardennes to encircle Belgium and trap the Allied armies. A month to the day from Tizard's urgency note, the position was disastrous; a huge supply of rich ore of the most deadly element was in German hands.

It was, also, far too late to make representations to Edgar Sengier, who had set up his base in New York, with a new company subsidiary he called African Metals. His reasons for leaving Brussels must remain a mystery at least until his papers are opened to public scrutiny. He died in 1965 and, in his will, placed a 50-year moratorium on his records.

Left in control of the ores at Ooolen, was a senior executive, Gaston André, who received terse and belated orders to dispatch 70 tons of refined ores by rail to Toulouse and to load the remainder in drums and ship it from Antwerp to New York.

André was not told of the urgency, nor did he know the high strategic potential of this industrial waste heap; there were other immediate demands of war. However, the loading had just begun when the blitzkrieg struck Belgium at 4:30 A.M. on the morning of May 10, 1940. It was then far too late for any "measures" Sir Henry Tizard could urge.

• 7 •

Denmark was about to be occupied. Very soon the world outside would be cut off by a ring of steel; there was perhaps an hour now left to get a message out. Niels Bohr had an irresistible urge to send a message as a valedictory gesture. He addressed it to his former Jewish colleague, Dr. Otto Frisch, in Birmingham; but, also, he thought of his friend, John Cockcroft, and the young English girl who had been tutoring his children; and he included in his message to Frisch, "Tell Cockcroft and Maud Ray, Kent."

In London, Sir Henry Tizard had formed a subcommittee to his main Committee for the Scientific Survey of Air Warfare. Its members were Thomson, Oliphant, Chadwick, and Philip Moon; its purpose was to deal with the mettlesome problem of uranium. It was deemed so important that no secretary was provided; Sir George Thomson was chairman and took the minutes in longhand. The importance of the group increased suddenly—with the impact of the Frisch-Peierls memorandum, the news from Paris, as well as hints through neutral channels that in Germany a Colonel Erich Schumann, a physicist with the High Command ordnance section, had taken overall control of the Kaiser Wilhelm Institute in Berlin. It was also reported that a Professor Paul Harteck of the Institute was seeking a Clusius tube for separation of the 235 isotope of uranium.

Then came Bohr's telegram from Denmark. Puzzled, Frisch, who had never heard of the lady from Kent, handed it to John Cockcroft, who produced it at a meeting of the subcommittee, and they all pored over the perplexing text. Cockcroft was moved to suggest it might be an anagram with a hidden message,

"You see, *Maud Ray, Kent,* could be a slightly garbled mixture of letters which actually spell *Radium Taken.* Is Bohr trying to tell us something about the Germans being at his Institute?"

And yet—maybe it went much deeper! The letter *U* most certainly would mean uranium ... then, would *D* mean disintegration? Would the whole word *Maud* imply there was German interest in "Military Application Uranium Disintegration"? If so, this was intelligence of the highest order. It meant the Germans were building an atomic bomb!

The stark reasoning from Bohr's cable was a stunning shock. The prospect of bombs weighing only a few pounds, destroying whole cities, had instant new currency. The impact was increased by other ominous rumors out of Berlin of Harteck's trying to separate uranium isotopes—secret meetings of leading physicists—reports of uranium going to the Kaiser Wilhelm Institute—the banning of the export of uranium ores from the old Joachimsthal mine, in what was now German Czechoslovakia.

The subcommittee was given a secretary from the Air Ministry and now became an advisory and investigative entity. Its real objective was the possible military use of uranium fission—to put Britain ahead of the Germans. It faced a daunting task in secrecy, winning much more money and getting a supply of raw materials from war-bound departments, all without telling what it was wanted for. And it made an historic leap; it took uranium from its place as an element of academic curiosity and put it into the role of a potential weapon of war.

This group of five—later to be added to—faced their secret

mission. They needed a code name to cloak their task in ano-
nymity. At Cockcroft's suggestion, believing they knew the awesome import of Bohr's clipped message, they called themselves—
The M.A.U.D. Committee.

Not until Niels Bohr was snatched to safety, three years later,
was it known that Maud Ray of Kent was a real person and not a
garbled anagram!

With the German net closed around Denmark, Norway enslaved, and attacks on the Lowlands, Frisch was utterly grateful
for the sanctuary of Oliphant's laboratory and for the shared
comfort of Peierls' home, at Edgbaston, near Birmingham.

Then the policeman knocked on the door. He quietly agreed
to go to the local station. He had heard of difficulties faced by
other exiles in Britain. He knew people had been interned, locked
behind wire for the duration of the war; and he believed these
decisions depended on whether wives or children would suffer.
But he was alone in England.

At the police station he was interrogated about his background, his time in Germany, his relatives, but not about his
work at the university. He said, later, "They quizzed me on so
many points, I could see it all added up to the question—'Is there
any reason not to intern this guy?' Then, they let me go."

At the university, because Oliphant was away, he spoke to
Dr. Moon, and the police were told of the importance of Frisch's
research. He was not openly bothered again and was unaware of
any surveillance of his mail going out of the country. He was to
learn nothing of the high level of interest; he did not know Sir
Henry Tizard had instructed the Director of Intelligence of the
Air Ministry, Air Commodore Boyle, that Otto Frisch should be
investigated as a security risk. For more than 30 years the information laid against Frisch by Allier, as agent of Frederic Joliot-
Curie, and the resultant action and recording of the interrogation
of Frisch as an enemy alien, were concealed in the secrecy of
Cabinet documents. By the time the facts were revealed, with
public access to these papers, the war was long gone, Joliot-Curie
was dead, and the secret uranium bomb no more than a thermonuclear detonator.

The police interrogation of Frisch caused only a pause in his
search for the path to the atomic explosive. He wrote of these

days, "One of my first needs was a measurement, however rough, of the fission cross-section of uranium-235 for neutrons similar in energy to those emitted during fission."

As he had done in the original physical proof of uranium fission, he prepared a small ionization chamber lined with natural uranium; he argued that if Bohr was right, if it was only the 235 isotope that fissioned, then he would register no reaction from the effect of neutrons on the uranium-238 atoms. He needed a powerful source of neutrons, and he heard there was a source of radon, a decay element of radium, owned by the Manchester Hospital. Wrapped in a coat of beryllium, this highly active substance, element 86 in the table, would be productive of the neutrons he wanted. However, there was a snag; it was kept, with the usual forest of glass apparatus for the separation from radium used for therapy, at the bottom of the Blue John Cave in Derbyshire. The tiny glass capillary could only be borrowed; and radon had a half-life of only four days.

He talked to Mark Oliphant, who asked one of his students to bend his gift for electronics to devise a means of detecting the fission pulses. The student's name was Ernest Titterton, and his talent for electronics was to take him to the firing of the first nuclear explosion.

Frisch went to Blue John Cave, a tedious train journey, and climbed into the deep cave on wet, slippery rungs of iron ladders. He returned to Birmingham with the fragile glass tube in a lead bottle in his briefcase.

At Birmingham, Titterton worked with him non-stop for 36 hours, and two noteworthy things emerged from it. Frisch obtained the fission capture measurements of neutrons he needed; and Titterton made a discovery for which he could never be credited.

It was Titterton's discovery that, at first, confused Frisch's results. The equipment persisted in recording small bursts of nuclear disruption, even when the radon-beryllium neutron source was removed. Frisch suspected the equipment was faulty; Titterton countered this by showing a new uranium phenomenon. True, uranium-235 fissioned on the intake of a neutron, and natural uranium-238 did not; yet the 238 atom also exploded, but entirely as a process of its nature. Rutherford's original observation of spontaneous disintegration could be applied now to the common uranium nucleus. Titterton had recorded the first step in the reduction of uranium-238, which goes through a chain of decay steps that finally reduce uranium to lead. Fate was unfair to him; this was secret work, and he could not publish his results. Two months later Russian researchers made the same finding and gained the credit. Igor Kurchatov, so-called "father" of the Soviet atom bomb, was one of the Russians. He daringly despatched a telegram about his triumph to the American *Physical Review,* and so was acknowledged as one of the discoverers of spontaneous fission in uranium-238. His message was published on July 1, 1940; it was clear indication for Western physicists that the Soviets were quietly working along similar lines.

Spontaneous fission did not detract from accurate estimates of 235 fission; rather it gave the Peierls-Frisch calculations greater precision—and this was to have its effect on the history of atomic energy. It produced immediately the need for a source of uranium mixed with fluorine, known as hexafluoride. Oliphant told them he thought Chadwick, at Liverpool University, might help, and so they went to see him. Frisch described it.

"I always remember Peierls and I waiting in his office, not knowing what to expect. Then, when he came in, hiding his kindness behind a gruff manner, he knew at once what we were seeking and asked how much of the hexafluoride we needed."

This meeting had a double outcome. It led to Frisch moving his base from Birmingham to Liverpool, mainly because there was a cyclotron there he could use in his drive for separate uranium isotopes. Also, with the gasified uranium Chadwick provided, they found that the Clusius tube would not do the job. This led to Peierls joining forces with yet another German exile, at Oxford, in a study of the separation of 235 by passing gaseous uranium through porous membranes. This was the gaseous-diffusion method, eventually to be used in huge atomic monoliths in America.

Peierls' new colleague was Francis Simon (later, Sir Francis); he had just been naturalized as a British subject. The combined talents of Frisch, Peierls, and Simon were critical to the target at which the M.A.U.D. Committee was aiming. As aliens, however, they could not then be added to the strengh of the committee seeking a British atomic bomb.

· 8 ·

War was at the very gates of Paris. The Joliot-Curies decided to escape to the south, to Clermont-Ferrand in the Massif Central. Frédéric and the now ailing Irène drove in silence in their Peugeot 402. Acrid black smoke clouded the sky, and then came the crump of bombs in the distance.

For Irène, there was a tearful flash of memory: She was with her mother when Marie was serving at the end of World War I with a radiography unit, and she had cried at the awful spectacle of the shell-mangled towns and villages. Marie had counseled her, "Don't be afraid, child. Remember, always, reach for calmness, try to be courageous. . . ."

The anger in Joliot-Curie had been a fierce pain from the day Raoul Dautry had telephoned him, "The Nazis have smashed through our lines at Sedan—tank columns are racing west to Paris to deal a death blow to the nation. . . ."

Kowarski and von Halban—released from their island captivity once the heavy water was in France—were already in the south with the dismantled apparatus from the Collège and the essential notes of their uranium work. They were installed at the Villa Clair Logis, a big house von Halban had rented in the southern suburbs of the city of Puy de Dome. The mysterious Product Z itself was still hunted. Allier had arranged for it to be safe in the vault of the bank's branch at Clermont Ferrand; but later, the manager had trembled in case the Germans found he had aided in the smuggling of a critical material. "Move it, at once," he had ordered, and, so, it had been taken to a new security—it now resided in the condemned cell at Riom Prison.

The 250-mile journey to Clermont Ferrand was an agonizing crawl. The road was under military patrol and there were columns of refugees from Paris making their way to the south coast. At times, the young assistant from the Collège driving the Peugeot pulled to the roadside when Stukas ranged overhead.

Joliot-Curie's dream of the balloon reactor was slowly disintegrating. Certainly, French leadership in uranium research would vanish quickly, and, with crushing defeat staring France in the face, he feared it could be the end of the Curie epic.

For two nights and three days, they were on the road, and the inhumanity of it all gave rise to a philosophy that would blossom in Frèdèric in the hard years under Nazi occupation. During that time Germans were placed in his laboratory, and they used his hand-built cyclotron; Langevin was jailed by the Gestapo; he himself was twice arrested and questioned. Finally, he was accused of collaboration by his own people when all the time he had secretly risked his life as a leader in the Resistance, managing underground laboratories to make plastique explosives and bombs, timing equipment for detonators, and Molotov cocktails for use at the time of liberation. He would emerge from those four endless years of terror and brutality an avowed Communist, bent on world peace.

The crisis for the Joliot-Curies expanded when they reached the Villa Clair Logis; soon Frédéric was faced with the gravest decision of his career. There was only a day or two to recover from the journey, a few hours of talking with von Halban, Kowarski, Henri Moureu—his deputy at the Collège who came south with the heavy water—and Irène, but she was mostly indisposed.

On the afternoon of June 16, walking with Moureu on the gorse-splashed hills of the volcanic Puy de Dome, Frédéric saw a car racing toward them; it stopped sharply and out jumped the secret service agent, Lieutenant Allier. Standing in the free air in the warm June sun, Juliot-Curie heard the ending of uranium work on French soil.

Allier brought orders from the minister for armament. He had been at a meeting with Dautry and the legendary Earl of Suffolk, the British nobleman-scientist attached to the Ministry of Supply, who was a special attaché to the Paris embassy. It had been arranged that as many important scientists as possible should escape to England to continue to work toward final victory. Joliot-Curie's orders were to repossess the heavy water and report, with his team, all his equipment, radioactive sources, all vital papers and records, to Dautry's *chef de cabinet,* a Captain Bichelonne, who had set up new headquarters in a school at Bordeaux. The Earl of Suffolk was making arrangements and would meet them there as soon as possible, for Bordeaux was under increasing air attack.

Allier then went to the Villa Clair Logis to enlist von Halban's aid to repossess the heavy water. They took the big Peugeot stationwagon and set off for the prison. At first, the warden refused to see them, and Allier made loud demands in the name of France; when they came face to face, the warden said he felt he would be risking his life to let the mysterious material out of his possession. He might have to answer to the new masters when they came.

Von Halban was furious, but Allier pushed him aside, leaned across the warden's desk and pointed his revolver at the man's head. He ordered, "I tell you now, you will give the order for the cans to be loaded into our wagon—or you will not live to know who your new masters will be! You will die—an enemy of France!"

Allier stood over the warden with the revolver while 20 convicts loaded the cans into the vehicle. Before leaving, Allier told the man, "We shall remember you, monsieur."

There was sad finality that evening at the villa; again loading the vehicles on the gravel-strewn driveway, Joliot-Curie spoke to his two research associates with formality, and separately. It was his farewell for the duration of the war. At dawn they drove in convoy, the big stationwagon leading, the young assistant driving with Kowarski's wife Dora beside him holding their injured four-year-old daughter, Irene. The child had been burned on an arm during the upheaval of escaping from Paris. (She was named after the heroine in *The Forsyte Saga,* and not for Madame Joliot-Curie.) Kowarski spread his 240-pound bulk across the packed cans and lead pots of radioactive sources in the back. Von Halban, his wife and daughter rode in the second car; Frédéric and Irène, with Henri Moureu, were in the third vehicle.

They faced a nightmare journey of 200 miles to the west coast port. Crossing many roads running south—all crammed with fleeing vehicles and patrolled by aggressive mobile military police—they stopped at times so Kowarski could crawl from the cramping torture of the low cabin and be sick at the side of the road. In the chaos they were separated from the last vehicle. They were not to know what happened to Joliot-Curie and Irène until the war was ended.

Nineteen hours after leaving the villa, the two vehicles containing the von Halban and Kowarski families drove into the schoolyard where Dautry had set up his office. Von Halban at once assumed command and went in to ask for their orders. Captain Bichelonne had them ready: papers of instruction and documents for them to sign accepting the conditions under which they would leave France on a British-commandeered ship, the *SS Broompark*. The vessel was waiting at the harbor docks with the Earl of Suffolk already aboard with other scientists, a wealth of diamonds, valuable machine tools, and blueprints culled from French organizations. Joliot-Curie would be directed to the ship when he arrived, they were told.

It was then past midnight; they went straight to the dock to report to the earl. Bichelonne watched them drive away; in front of this ambitious, slick young man was a brief career as a minister in the Vichy government, the stigma of collaborationist, and, finally, death in Germany.

The *Broompark* was a dirty, cramped old collier. Dozens of men lolled on tarpaulins covering piles of coal dust; the few cabins were jammed with the women and children of the fleeing scientists and other important people. Yet one man stood out from the rest. Barechested, his body marked with many tattoos, giving orders, swearing in voluble French with an atrocious accent, this was the twentieth Earl of Suffolk, Michael John James George Robert Howard, Baron Howard of Charlton, Wiltshire, one of whose ancestors was Howard of Effingham, Admiral of the Fleet to Queen Elizabeth when Drake defeated the Spanish Armada.

With a burly bodyguard, ex-sailor Fred Hards, the earl was cutting wood to make a roomy raft and cursing that there were no nails and it would have to be fastened with ropes. When the cars arrived, he had sailors bring the heavy water cans and lead pots to the bridge for binding to the raft. Then he called for a British officer, Major Golding, and, with their hands on the Bible, Kowarski and von Halban had to swear that, if the ship was struck and sinking, they would stay with the raft and float from the bridge; they would not try to aid the women and children. Suffolk told them, "On that raft are things more precious than any of us—likewise your brains and knowledge. If disaster strikes this ship you are pledged to stay with the raft until you are picked up—or you are dead. Now, you can sleep on deck."

There was no rest for anyone that night. The sound of bombers was heard, incendiaries rained down, and the docks were soon blazing. While Kowarski and von Halban scanned the shore for sign of their chief, the captain eased the vessel into the stream and sailed to open water, there to anchor in pitch darkness.

Von Halban again took command. He told Suffolk, "We would like to go ashore to search for Professor Joliot-Curie and his wife."

The Earl of Suffolk, who had sailed before the mast around the Horn, who had been a jackeroo in the Australian outback, patted the shoulder of the ex-German scientist.

"Not for me to say, old chap. You ask the captain, but I know he'll tell you that nobody goes ashore—once aboard. I hope Joliot-Curie will come aboard to see you. And, if that happens, I doubt whether he'll be allowed ashore again."

The ship swung at anchor in the dark while Bordeaux flamed. An hour after his conversation with von Halban, the earl and Fred Hards quietly pulled away in a rowboat.

On the harrowing car journey westward, Irène Joliot-Curie became increasingly ill. Ailing from the radiation damage that would kill her, she reached the point where she felt unable to travel further. At Clairville, in the Dordogne, they diverted to find a hospital where Irène could be treated. Distraught, Joliot-Curie drove on to Bordeaux with Moureu.

They reached the schoolhouse after two in the morning; on the stairs, with Bichelonne, they met the Earl of Suffolk. They had met many times during Suffolk's term of duty in Paris; now, the Englishman threw his arms wide in welcome.

"Come, cheer up, Frédéric! All is not lost yet. Come with me. You've just time to get aboard with your two associates before we sail. I'll personally see that Irène and the children are smuggled out of France. We'll send a destroyer to pick them up with other people who are waiting on the west coast. Come, hurry! You can continue your work in the safety of England."

Joliot-Curie could not take the step. He could not face sailing away, leaving Irène ill and his children in the north. He could not run away—like the politicians who had betrayed the people. Face grim, dark eyes very sad, he took Henri Moureu by the arm, and they walked away, leaving the earl and Bichelonne on the stairs.

· 9 ·

Edgar Sengier, from his room in the offices of the African Metals Corporation, could look down on sunlit 42nd Street in New York that late May day, 1940, to scenes of a busy, vital city. In his mind's eye was a picture of his homeland, crushed by the most powerfully mechanized army the world had known. Fears he had held of the total annihilation of Belgian defense were confirmed. In a matter of days, the tide of Krupp steel had scattered and broken the armies of France, Holland, Belgium— and Britain. Indiscriminate, ruthless bombing, fire and destruction, had swept towns and villages in the path of German advance. Antwerp, Amsterdam, Brussels, Copenhagen, and Oslo echoed with the tramp of Nazi boots; and soon they would be heard in Paris. The Wehrmacht looked invincible; but it was the speed of the conquest, and its ramifications, that were so appalling.

The wealth of the great company he controlled, the stocks of copper, cobalt, silver, the uranium ores in Belgium were lost; and he was helpless to aid his staff and workers now inside the steel ring.

The immediate question was—did the Germans know the hidden value of the yellow-cake waste heaps at Oolen? Was the potential known to Hitler's scientists—the men who had sought the heavy water in Norway?

What would stop them from doing what the French had in mind? Why couldn't the Germans, just as easily, build an atomic bomb center in the Sahara? Once there, with their lightning speed of strike, how long would it take for a task force to sweep into the Congo and capture the biggest uranium source on the planet?

It could not be risked. There could be no delay. He sent orders that same day to the Union Minière representatives at Elizabethville in the Congo. All uranium ore at Shinkolobwe must be packed into steel drums, quietly and without attracting attention. Whatever ruse or cover story necessary should be used to get the thousands of steel drums on the the snaking old trans-African railway for the 1,200-mile journey over the uplands, through Portugese Angola, to the Atlantic port of Lobito by early August.

Sengier then scouted the market seeking a charter freighter that could pick up the drums at Lobito and bring them across the Atlantic to the docks at Port Richmond on Staten Island.

He had seen no sign of official interest in uranium in America—as there had been in London and in Paris—but he felt that some of the ore had to be moved as far as possible to a place of safety.

The *Broompark* emerged through the channel mists, darkly; in the early hours of June 21, 1940, she dropped anchor in Falmouth Harbor. For the second time, the world stock of industrial heavy water came to England.

Next day, a London taxi halted at the doors to the sandbagged offices of the Ministry of Supply; it stood with its meter ticking over, and the Earl of Suffolk—in bedraggled greatcoat and dirty gum boots—strode into the building for a brief meeting with the under-secretary to the department, Mr. Harold Macmillan.

The meeting stayed vividly in the memory of the future prime minister; he saw a man with sleep-hungry eyes, his face blackened with coal dust, but shining with dignity and distinguished grace. The earl "was quick to state his purpose," to relate the mission entrusted by Herbert Morrison, the minister, and the task achieved with his French counterpart, Dautry, and to say there were valuable people still on the coast, near Bordeaux, waiting for a British destroyer. He told Macmillan of delays at Falmouth, that he had installed his precious charges for the night in a hostel and brought them to London by special train that morning. The buccaneer earl then said he had a taxi waiting, with about £5 million in diamonds, two atomic scientists—and "something called heavy water."

They talked, made arrangements, and the earl went off to deposit his diamonds at the Bank of England, the two scientists at the Great Western Hotel, and the heavy water cans temporarily—for the second time—at a prison, Wormwood Scrubs. London was liable, in those weeks after Dunkirk, to heavy air attack, and shortly afterward, the persuasive earl had them lodged in the vaults at Windsor Castle, under the Royal Library.

The product Allier had snatched from the Germans in Norway was safe in England, waiting for an historic meeting with uranium oxide. In the Great Western Hotel, with the papers brought by Kowarski, was Joliot-Curie's blueprint graph of the atomic balloon.

In later years, Macmillan recalled that brief meeting with Suffolk: "I have never known in a single man such a remarkable combination of courage, expert knowledge, and indefinable charm. He was a true Elizabethan character."

On returning from France, Suffolk at once took a liking to defusing parachute mines and unexploded bombs that came with the German blitz on England. Less than a year later, on May 12, 1941—with his ever-faithful Fred Hards and his fearless blonde secretary, Eileen Morden—he was trying to disarm a bomb, dropped six months before, when it exploded, destroying all three.

· 10 ·

At dawn on Friday, June 14, 1940, the Germans occupied Paris.

Among the leading units was a specialist group, similar to that which would be employed by America when France was liberated. This German unit, backed by assault troops, was headed by a physicist, the same Walther Bothe whose work in 1930, with Becker, had led the Joliot-Curies to provide Chadwick with the clue to the neutron. Bothe was now a leading researcher at the Kaiser Wilhelm Institute, and his present mission was to strip Joliot-Curie's laboratory at the Collège de France of its secrets and its precious materials for the benefit of the Third Reich.

Finding the Collège laboratory locked, Bothe had the doors forced and, at once, personally searched for the heavy water, for uranium, for radioactive sources; and he rifled papers for traces of the lines of uranium research. He went away empty-handed.

When Joliot-Curie returned to Paris from Clairville, where he had waited for Irène to recover sufficiently to travel, he found an eerie situaiton: a half-empty "German" Paris, a deserted laboratory.

Shortly after he tried to take up routine life, the top man behind the Nazi uranium effort came to the laboratory. General Erich Schumann was overseer of uranium materials captured in

Belgium; he had set the patriotic Werner Heisenberg—winner of the Nobel Prize for Physics in 1932—at the head of a group in the Kaiser Wilhelm Institute to develop a German uranium reactor. Schumann prided himself on his knowledge of physics; he was scientific adviser to Field Marshal Keitel and served on the Third Reich Research Council.

With this background, he greeted the French scientist as a colleague. In a scene that was "pure seduction" in Joliot-Curie's eyes, Schumann paid respects to a fellow scientist "with a sparkling reputation," and said it would be a privilege to have him work side by side with German scientists. With Schumann was another German physicist, Wolfgang Gentner, who had studied in Paris in the 1930s; he was known to have anti-Nazi views and was to be accused by the Gestapo of having "democratic" ideals.

Schumann was taken aback when Joliot-Curie said, bluntly, neither he nor his laboratory would be used to further the German war effort, that his only interest was in pure academic investigation.

The Nazi's bland approach altered.

"We know what you have been doing here, of course. Burning or removing all your papers was of no use. You see, we have captured copies of all your reports to the armaments minister—in a railway carriage at La Charité-sur-Loire. Now, tell me, Professor, where have you hidden the heavy water?"

"It has left Bordeaux by ship for England."

"And the uranium ores?"

"They, too, have been taken away by the ministry—to some location I do not know."

Schumann did not believe him. He told Gentner—who later related it to Joliot-Curie—he believed the ore had been shipped to North Africa. "I think that is where we should send a mission to search for the uranium and the heavy water."

The uranium was not critical in Schumann's plans; big stocks of Congo ores were being trucked from Belgium to Germany, and the heavy-water plant in Norway was under command to step up production. The Germans lacked nothing in the way of materials; their critical loss was the talent and insight of the handful of exiled scientists who were, even then, shaping the pattern of nuclear weaponry.

In the summer of 1940, two of these scientists, Peierls and Frisch, wrote a ten-page summary that was the most up-to-date understanding of uranium's potential. Nine separate treatises sketched outlines of awesome possibilities. It was detailed extension of their memorandum in March, on which they had heard not a word in reply from the M.A.U.D. Committee. Worried that German work on uranium could be leaping ahead, they rounded out the theoretical picture:

A chain reaction in natural uranium was out of the question for achieving an explosion; it might serve as a means of energy, but not for a bomb.

With constants and estimates, reached from Frisch's work with gasified uranium and the neutron source from the Blue John Cave in Derbyshire, they could say a chain reaction in 235 was possible and, if achieved, would cause a release of energy quite suddenly, in a fraction of a second.

The hailstorm of neutrons at the heart of a sphere of 235 metal, the unleashed energy and heat from only a single pound in the center, would create an explosion equal to many thousands of tons of TNT.

And, with the fearful heat, with the great shock waves, there would be the *lethal radiation of fission by-products.*

Their calculations put the weight of 235 metal needed for cataclysmic eruption at around 8 kilograms. They reasoned that if this sphere of uranium was wrapped in some other metal, to reflect escaping neutrons back into the critical mass, then it would be possible to reduce the size of the 235 metal to around five kilograms.

The Peierls-Frisch papers, however, presented massive problems for reaching this goal; there were issues of stray neutrons causing pre-fission so that the explosion was premature and slowed down, like a wet firecracker; there was difficulty in bringing two pieces of 235 metal together, two subcritical sizes that had to move at lightning speed to form a critical mass—at some 6,000 feet per second. Also, there was the very formidable problem of how 235 metal was to be obtained.

They said 235 atoms had to be winnowed from the normal 238 atoms; they were chemically identical, only slightly different in weight, and *very, very* slightly smaller. How could they be segregated efficiently with speed and economy?

Peierls and Frisch probed deeply into this problem. Thermal diffusion with the Clusius tube made a dismal showing, as did centrifuges and evaporation—so they decided on gaseous diffusion. They reasoned this was the best prospect; they developed fifty or so separate formulas and produced a process for passing gasified uranium through the smallest pores that could be made in a metal screen, holes tiny enough to make it easier for the smaller 235 atoms to pass through than for the 238 atoms. They took this idea through its stages; pressure pumps to move the gas, power consumption, time factors. They found the gas would have to go through some 3,000 phases; they concluded this process was not so slow as it seemed, that it would be a matter only of days.

They completed the ten-page summary preview of technique that would eventually be used at the huge gaseous diffusion plant at America's Oak Ridge. They drew the first blueprint for the type of bomb that would shatter Hiroshima. And they were still two exiles, restricted in their freedom, barred from working on what ministry officials called "security projects."

Peierls handed the papers to Sir George Thompson; the M.A.U.D. group met in secret at the Royal Society headquarters and minutely scoured the equations and the reasoning for errors. Everything stood up; so they agreed to launch a pilot study and gave the gaseous diffusion task to Francis Simon, who headed a team at the Clarendon Laboratory in Oxford.

In the Cavendish Laboratory in Cambridge, where he had proved the existence of the neutron, James Chadwick watched the Joliot-Curie atomic balloon rotate on its spindle.

All through autumn—it was now November—Kowarski and von Halban had cooperated to assemble two semi-spheres of spun steel fabricated at a Birmingham plant. Kowarski remembered, "At times, we used a Meccano set and cellotape to hold pieces together, and, in the end, it worked."

Inside the two-foot-diameter sphere, the paddle stirred two kilograms of uranium oxide, distributed it through the heavy water—retrieved from Windsor Castle in midsummer—and the neutron source fired its atomic projectiles. And more, many more, neutrons emerged than went in.

It produced no power; it gave off no heat. It proved what the cautious called a "divergent" chain reaction; it meant that out of Joliot-Curie's vision had come the first empirical proof. With the right moderator, with much more uranium, with controls and the right assembly, man had a new source of energy. Two years later, in Chicago, this vision was totally manifested.

Chadwick, containing his excitement, told the M.A.U.D. Committee it was "overwhelming proof." Chain reaction was a reality; all that was needed was management.

At that time the brains in M.A.U.D. were geared not to the work of slow neutrons for energy release, but to fast neutrons for bombs. The atomic balloon was interesting, but a long-term project for peacetime.

In the first weeks after their landing at Falmouth—when Kowarski and von Halban had been put on the staff of the Ministry of Aircraft Production, at a guinea a day—the two scientists were asked to write a report on the Paris fission studies. They concluded that "there may be two processes in the fission of uranium. As well as neutrons rupturing the structure of 235 atoms, the normal 238 atoms could capture—and absorb—one neutron each. This may give rise to spontaneous transformation; the result may be a new kind of isotope which could fission as readily as 235. . . . ."

By this time James Chadwick had made a calculation that put the volume of the critical mass of uranium-235 at 60 cubic inches. The 235 metal had yet to be made, but this denoted a sphere between five and six inches in diameter. It was while he was doing this work that Chadwick was astonished to see a report, by McMillan and Ableson of Berkeley, California, published in the American journal *Physical Review,* which announced the discovery of such a neutron-induced uranium by-product as Kowarski and von Halban had suggested in their secret report. A classic finding, the first man-made step beyond element 92, uranium, it had revealed a transuranic substance, element 93.

It started, the Americans reported, as uranium 239—the common uranium 238 plus a neutron. The half-life of this new element was two days and eight hours; then it threw out a beta particle and was transformed into element 93. What was far more significant—there were indications the new element then changed to *something else.*

Theory said this *something else* was element 94!

It was a prospect of momentous consequence. In Cambridge, this same possibility had been reconnoitered. The M.A.U.D. group had taken their reasoning on irradiation of uranium to the prospect of this synthetic element. Dr. Egon Bretscher and Dr. Norman Feather even suggested that element 94—if it existed—could be much like uranium-235, perhaps even *more* fissile. If so, a smaller volume would be needed to form a critical mass.

It was still only theory, but it was red hot. Uranium could produce a most dangerous daughter product, a super-super explosive!

They also talked about naming these irradiation products. One of the team, Dr. N. Kemmer, said the use of planet names, started by Klaproth, should be continued. Outward from Uranus was Neptune; element 93 should be named Neptunium. The next planet was Pluto—and element 94 should be known as Plutonium. It was exactly the reasoning to be followed in California.

James Chadwick, however, was hightly incensed with the California disclosure; it was foolish and inconsiderate. Britain was engaged in a deadly race with the Germans—and here were the Americans handing them a momentous find on a plate—in print.

He asked Cockcroft to have a diplomatic protest fired off to Washington—and, indeed, the workers at Berkeley decided voluntarily to impose a code of secrecy, with the magazine editor cooperating by accepting and dating, but withholding the printing of their reports, so they should not lose the credit of being first. It was gracious and wise since America was still not in the war and researchers were not under the same threats and pressures of those in Britain. For American scientists—apart from the few European exiles—interest in uranium fission centered on slowed neutrons for energy release and the arcane structures of matter, not on prospects of new weaponry.

In September, 1940, Sir Henry Tizard flew the Atlantic to arrange an exchange of data with American science. It was to be as far-reaching as it was one-sided.

Tizard took with him what was then known as the "precious black box," containing blueprints and data on new weapon systems and inventions. He also took two brilliant men—Dr. John Cockcroft, who had split the atom in 1932, and Dr. E. G. (Taffy) Bowen, virtual inventor of radar.

Kenneth Bainbridge, when professor of physics at Harvard, was to say that "E.G. Bowen started things off with the magnetron, indeed, with the whole outline of components, systems, and military uses. . . . It was one of the great inventions of the war."

Tizard and Bowen carried with them a highly powered magnetron, which had been developed by Oliphant and associates at Birmingham—H. A. Boot, J. Randall, J. Sayers, and Ernest Titterton.

Radar was Tizard's long suit; the uranium question seemed almost a side issue and was left to Cockcroft and Ralph Fowler, who came down from Ottawa to join the Washington talks.

A consequence of this visit was that the British government accepted Tizard's recommendation to set up a scientific data exchange office in Washington; the U.S. National Defense Research Council was invited to set up a similar office in London. Through these channels it was planned that data would flow back and forth across the Atlantic.

In practice, the arrangement proved to be almost a one-way affair; British groups soon grumbled that, while their research results poured freely into America, very little came back. Through the rest of that winter, and the spring of 1941, the lack of response aroused increasing anxiety. This climate was to affect future Whitehall decisions on the American offer of a joint uranium project.

The dissatisfaction among the scientists did not stop the flow of data to America; in the summer of that year, through the channels which Tizard had set up, one of the most momentous documents of World War II was despatched from London.

· 11 ·

The discovery of uranium fission was not yet two years old when the technical means to use it as an explosive came into the hands of the M.A.U.D. Committee. In their paper, "Estimate of the Size of an Actual Separation Plant," Francis Simon and his colleagues gave solid basis to the theoretical attainment of Peierls and Frisch. Modestly, they said the idea of diffusing gas through porous membranes was not new; a German physicist, Hertz, had built a small plant to do this ten years before (and, again, the fear of Germany having the means to race toward the bomb was heightened). Simon and his associates, however, had tackled the prospect on scale. They reported on the following:

> A complex able to deal with enormous amounts of gasified uranium in which very high purity 235 material could be won from comparatively simple equipment. Purity—higher than had been dreamed of, a staggering 99 percent—could be gained in the short time of 12 days.
>
> It meant a vast area of 70,000 square meters of porous metal membranes. There would have to be 18,000 units, with each unit having 20 stages; and the whole plant would cover about 40 acres of ground and would need a supply of 60 megawatts of power to operate 70,000 tons of machinery. They put the construction cost at £5-million, the annual operating cost at £1.5-million, and said this complex could produce a single kilogram of uranium isotope-235 each day.
>
> It would take about 18 months to build.

Francis Simon and his team concluded that "separation (of 235) can be performed in the way described, and we even believe the scheme, in view of its object, not unduly expensive of time, money, and effort."

Britain, facing the mightiest army in history across the Channel alone, with catastrophe looming in the East, and the desert forces in North Africa still two years from Churchill's "gleam of victory on their helmets," was at a watershed in history. And here, suddenly, was the way to a war-winning bomb.

The Oxford team's blueprint complex offered fissile material for three atomic bombs every month of the war. It raised the prospect of the Ruhr, Berlin, Hamburg devastated in the first four weeks, then Munich, Stuttgart, Bremen in the next month. And *not* by costly, high-casualty, massed bomber raids, but by a single aircraft flying high over the cities.

Again, hesitation and doubt intervened.

At the end of the first year of the British search for a way to the atomic bomb, the M.A.U.D. Committee had made remarkable advance, although hamstrung by poor resources, and hindered by bureaucrats in high places. By the spring of 1941, the Frisch-Peierls 235 memorandum had been fleshed out:

> The probable critical mass of the uranium isotope had been agreed on by the best nuclear brains in the land.
>
> The Oxford report showed the way for an atomic-armed Great Britain;
>
> The Joliot-Curie "balloon" in the Cavendish had signposted another road to a further nuclear explosive, perhaps easier to handle than uranium-235.

In Britain, the total effect of such a bomb on an enemy populace now swung into focus. The men of M.A.U.D. had known of this insidious aspect from the start; it came into open discussion with a special report to the War Cabinet's scientific advisory panel: "Neutron and gamma radiation, radioactive dust scattered into the air could be as deadly as blast and heat."

The expert consulted in this matter, Dr. L. G. Grimmet of the Medical Research Council, stated bluntly, "One single gram of uranium-235 suddenly disintegrated will give the same heating effect as an explosive charge of 10,000 kilograms of nitroglycerine."

The same amount of uranium-235—about one twenty-eighth of an ounce—would emit a radiation dose of neutrons and gamma waves equal to 20,000 million roentgens! As an expression of the biological effect of this on a human being, Dr. Grimmet said a dose of 10,000 roentgens would be fatal; 1,000 would produce serious effects; and only 100, administered to the whole body, would cause illness in a short time: "Any person at a distance of 50 meters from the fission of one gram of uranium-235 would receive fatal blood damage and would die of anaemia, toxaemia, and inanition. The dust from the uranium explosion would be highly radioactive . . . and it is well-known that radium dust, inhaled, or swallowed, becomes fixed in the skeleton and can cause blood changes and death."

The invisible threat in radium had come full circle since Pierre Curie expressed his fears in 1903; those same dangers were a further cogent reason to build and use the uranium bomb.

The M.A.U.D. group had known about these factors from the first; now, the War Cabinet knew the atomic bomb could go on destroying people after the fierce heat and the blast were gone. As yet—although scientists were aware that ionizing radiation caused growth changes in plants and insects—nothing was said of possible genetic effects on human beings.

· 12 ·

The M.A.U.D. group commenced its task of making a comprehensive report to the government in the spring of 1941. Essential data for guidelines to building a uranium bomb was in their hands; only precise measurements were missing that would have made the outcome conclusive to laymen. The expression, and the phrasing for nonscientists, was a difficult chore that took many meetings, many redrafts, until the precise minds of Chadwick, Cockcroft, and Oliphant were satisfied they had told the story as it was.

If a true cradle had to be chosen for the birth of the nuclear bomb, it should be the old Royal Society building, Burlington House, London, where the men of M.A.U.D.—likened, since, to nuclear midwives—met to deliver this atomic offspring to the world. In June, the first American scientist to sit in on a discussion on the M.A.U.D. report was Kenneth Bainbridge, the young student at the Cavendish, in 1934, to whom Rutherford had exploded about Szilard's idea of a chain reaction in the light elements. Other American observers came later—among them, Harold Urey, who had discovered the heavy hydrogen atoms, and George Pegram of Columbia University. These three Americans were told by the normally cautious Chadwick, "I only wish I could tell you the bomb is not going to work—but I'm 90 per cent certain that it will."

When the text of the report was finally agreed upon at the beginning of July, German panzers were already smashing into Russia, and Britain had an ally at last. But there were grave anxieties; if the Soviet was toppled in the mad race toward Moscow, with all the resources and slave labor that would then come under Hitler's control, what could be the prospect for Britain without a war-winning weapon? The M.A.U.D. group believed the uranium bomb could win the war.

The free world was in deadly danger in mid-1941; Britain was under nocturnal onslaught from savage bombing and incendiary attacks; for the British government, there could be no question of withdrawal on moral grounds. The uranium bomb could mean survival. Also, there was the known German interest in heavy water and uranium.

Even if Nazi Germany was defeated on the battlefield, it could reverse the situation with sole possession of the uranium bomb. The facts had to be faced. It was a brave stand to take, with a government scraping the national barrel for men and materials, to urge a new, costly project.

They opened the M.A.U.D. report thus: "We emphasize ... we entered this project with more scepticism than belief, though we felt it a matter to be investigated. As we proceeded, we became more, and more, convinced ... we have reached the conclusion it will be possible to make an effective uranium bomb, containing some 25 lbs. of active material, which would have equivalent explosive power to 1,800 tons of T.N.T. . . . and would release large quantities of radioactive substances that would make places near where the bomb exploded dangerous to human life for a long period. . . ."

They underrated the explosive power of the bomb by a factor of ten; also, they did not say that Joliot-Curie's rotating balloon in Cambridge opened the way to production of a new fissile material. In all else, however, their assessment forecast the pattern of future development with astonishing accuracy.

They drew guidelines:

Separation of uranium-235 from ordinary uranium can be achieved by a system of passing gas through porous membranes.

Preparation of this material into compacted metal increases its density.

A gun barrel assembly is possible in which two sub-critical pieces of metal can be fired into each other at about 6,000 feet per second, thus causing a fission explosion.

Estimated cost of the necessary complex is around £5 million; and, with an output of 1 kilogram of active material a day, three bombs would be readied each month at a cost . . . cheaper than conventional explosives.

Their opinion on the capital cost was that "in spite of the large expenditure, we consider the destructive effect, both material and moral, so great that every effort should be made to produce bombs of this kind." They made a further pertinent point on the danger that physicists in Germany could even then be working as secretly as the M.A.U.D. teams: "The Germans have taken much trouble to obtain heavy water; they have uranium, and the lines we have followed are likely to suggest themselves to any capable physicist. . . ."

The report went into channels, to the Director of Scientific Research at the Ministry of Aircraft Production, Dr. David Pye—who had been quietly worried at the M.A.U.D. spending of £12,000, which he would be hard put to explain. Pye sat on the report for a whole week; then he sent it to Sir Henry Tizard. Finally, it reached the minister, Colonel Moore-Brabazon, and it got to Lord Hankey, War Cabinet member, a month later.

· 13 ·

Professor Sir George Thomson, at Imperial College, found the delay and procrastination of bureaucracy "absolutely maddening." More than two years had passed since he had taken the train to Cambridge to discuss the Paris report on uranium chain reaction with Bragg. Pickthorn had then warned the government the uranium question was a matter of "hours, not even days"; Tizard had said it was "all greatly exaggerated." Now Thomson believed the bomb was within reach of technical effort. Again he caught a train to Cambridge and spoke with Sir Will Spens: "I know the report must go to Lord Hankey; would you again intercede and speak to him of its vital importance to national security?"

Spens wrote to Hankey, saying he could rely on Thomson's judgment: "Thomson is Professor of Physics at Imperial College; he is a Nobel Prize-winner and is one of the best—if not *the* best—opinions in this country on such a matter...."

Lord Hankey was already under pressure. Advocates from the powerful I.C.I. combine, seeking a commercial share in the development, were Lord McGowan and Lord Melchett, and Hankey lunched with them and Lord Cherwell, Churchill's science adviser. The lunch party of four peers did not go unnoticed; Hankey had a note from Tizard which said, "I hear you have been discussing M.A.U.D. in the highly optimistic atmosphere created by the enthusiasm of three well-known peers."

Tizard's doubt, inherent in the note, was even more apparent when Hankey called meetings of his special panel of scientific advisers. This eminent group was chaired by Hankey himself; the other five men chosen to decide whether Britain should be first to build the uranium bomb were Sir Henry Dale, aging president of the Royal Society, Professor A. V. Hill, and Sir Edward Mellanby—all three were specialists in medicine or biology; Professor Sir Alfred Egerton, a chemical technologist; and one physicist, Sir Edward Appleton. As witnesses they called Lord Cherwell, Chadwick, Blackett, R. H. Fowler, Dr. Pye, of the ministry, Dr. H. L. Guy, an engineering expert, and, of course, Sir Henry Tizard.

At the meeting which dealt with the matter of a pilot plant to separate 235 from ordinary uranium, Tizard voiced opposition; he said the construction of this pilot plant would make "considerable demands on the time of skilled engineers experienced in the design of turbines." Also, he felt anxious "at the possibility that this might interfere with the progress of certain highly important developments now taking place in the design of aircraft engines which, as compared with the proposal for the uranium bomb, give more immediate promise of successful results...."

When Dr. Guy spoke of the vulnerability of a 235 separation plant to interruptions from air attack, the panel moved one step down the path toward transplanting the products of the M.A.U.D. accomplishments overseas.

The panel's recommendations to the War Cabinet opened further the door to opting out of the decision whether or not to build the world's first uranium bomb in Britain: "If it ... is decided to build a full-scale separation plant ... priority will again arise, since the work will make demands not only upon engineering personnel and special types of machine tools, but also for materials and skilled labour, demands which are bound to be large. This consideration points to the desirability of constructing a full-scale plant in North America. . . ."

Chadwick, on the grounds of secrecy, was opposed to this move but could not sway the panel.

Dr. Guy's warning of air attack on a gas-filled plant gave strength to the trend to go to the vast resources in the U.S.A. There was not, then, in all Britain a man with the power, the vision, and the resolve to see that the first country with the weapon could decisively end the war and avert the danger of facing defeat by any other country that had determined to follow the same path. Nobody knew how long the war would last, nor the course it would take. Russia was fighting for its life, and America was still on the sidelines; opponents of the project argued it was unlikely it could be finished before the end of hostilities.

The panel's report went to Churchill, via Lord Cherwell. Churchill, ignorant in physics as were Hitler and Roosevelt, nonetheless saw its potential. He created a new division—which was code-named Tube Alloys—to deal with the issue. He asked Sir John Anderson, who had once studied physics, to direct the undertaking.

This was the kiss of death for M.A.U.D. Anderson enlisted two executives from I.C.I.—Wallace Akers and Michael Perrin—as chief and deputy chief, a move which seemed ludicrous to observers, inasmuch as the panel had turned down the idea of I.C.I. running the program in favor of a bureaucratic organization.

The men of M.A.U.D. knew nothing of the ponderous creaking of the government process. The weeks went past with the great issue of the uranium bomb hanging fire. Late in August, Thomson spoke with Cockcroft and Oliphant, airing concern. To him, all seemed in abeyance; despite the fact that all M.A.U.D. documents and the full report had gone to America, no response had been elicited. Oliphant then revealed he was about to make a crossing, by bomber, with details of the latest development on microwave radar reached in his Birmingham laboratory.

Thomson said, "It might be very helpful if you could look

into the reasons why we've heard nothing back from them on our reports. . . . You know all about it, and nobody else could do it better."

In the summer of 1941, Churchill asked his chiefs of staff for an opinion on whether or not to build a uranium bomb in Britain. He received their urging for prompt action with high priority. The new Tube Alloys Directorate had been set up within the Department of Scientific and Industrial Research and was now out of the influence of the Ministry of Aircraft Production; yet some opposition still existed against the atomic bomb plan.

Unaware of what was happening, the M.A.U.D. team in their meetings expressed anxiety over the unexplained delay and the nagging problem of where Britain was to obtain supplies of uranium ore to support any future project. When this matter of supply was raised with Sir John Anderson's advisers, at a time when there were only a few tons in Britain, it was blandly assumed that "uranium from the Congo, or from Canada, would be made available." Indeed, no more uranium ore was to come into British hands until the American bomb project was near completion.

The scientific advisory report to the War Cabinet finally went to Sir John Anderson in the last days of September. He decided to form yet another advisory council, and some of the original M.A.U.D. men were placed on this. It was all bureaucratic delay and buck-passing. The M.A.U.D. report, finished in early July, 1941, was still only a document under discussion. Some of the M.A.U.D. people were not informed of the passing over of M.A.U.D. in favor of Tube Alloys until they learned it in December in a letter thanking them for their past services.

Professor Mark Oliphant, on his way to America in September, knew nothing of this. He was to be very angry and would write to Sir Edward Appleton that he could see no reason whatever why the people put in charge of the uranium bomb project should be "commercial representatives, completely ignorant of the essential nuclear physics on which the whole thing was based. . . ." To Oliphant it was "organisation tantamount to that in the United States where the whole thing is in the hands of non-nuclear physicists and is, therefore, being badly mismanaged. . . ."

Oliphant had special reason for his anger. It was three years since the discovery of fission, and he had been instrumental in pushing forward the Frisch-Peierls memorandum which, with Allier's testimony, had "started the whole shebang."

Questing into America, he knew nothing of the readiness expressed by President Roosevelt's advisers to Charles Darwin to enter a joint project of bomb development with the British, who were then so far ahead; nor did he know that delay and disinclination to join a common bond with the Americans was to fritter away precious time and allow the magnificent equipment and the larger corps of scientists in America to make such rapid headway that, in the end, there was little left for the British to bargain with.

· 14 ·

Oliphant was to make 20 wartime trips to America, on secret government business, shunning the clipper service out of Lisbon as risky and time wasting, traveling instead by bombers.

Tizard sent him on this journey, in late August, 1941. The latest advance made by his team on the magnetron—heart of the microwave radar system that could detect moving aircraft beyond the curve of the earth—had to be delivered and explained.

As the propellers whirled, he mused: Radar, to detect and destroy; and uranium-235 to unleash stupendous and, hopefully, crippling blows on the enemy . . . a gesture of trust in this exchange of information—albeit to a noncombatant ally—with little coming back. Yet what did that matter as long as America with her great resources, splendid equipment, and plentiful researchers took steps to see the free world was not caught in the lethal predicament of facing dictators armed with nuclear weapons?

Oliphant had no reservations on the use of uranium bombs: A mere 10 kilos of 235 metal would produce an impact, Chadwick believed, such as the Siberian meteor or the disastrous ammunition ship explosion that wrecked the Nova Scotia town of Halifax in World War I. And he had a long association with the element.

It was in the Physics Department of Adelaide University, where he had been a research assistant in 1925, agog with the notion of transmutation of one element to another, that Oliphant had turned to the appeal of physics from biology; and he had mounted an X-ray tube to produce electrons to bombard a uranium sample. He disregarded the flood of X rays to which he was exposed in seeking to insert electrons into the structure of the uranium atoms. He reasoned if he could do that, then the nature of the matter might change. He found the oxide of uranium hard to keep in place; so he mixed it with a wax compound, and then—as Klaproth had done—baked it at a high temperature and got gray, steel-colored metal beads. He hammered these into the small disc which was to be the target for his electron beam. He sat for hours while the uranium button got hot. It changed nothing, other than help win him a place in the Cavendish with Rutherford. Now, after all these years, it had fallen to him to urge the Americans into action on uranium.

Washington was pleasant at first—a different world to bomb-blasted Britain. His radar device went into friendly hands at the office of the National Research and Development Committee, which the President had set up recently. From there he went to a meeting with the chairman of the Uranium Committee, Dr. Lyman J. Briggs, a soil scientist who, by being Director of the National Bureau of Standards, had fallen into this task. And when he met this aging, tired man, Oliphant was taken aback.

There were several members of the committee called together to talk with him, including a navy officer and an army officer. Briggs invited him to speak of how things had gone in Britain on the uranium work, and, as he did, he saw bewilderment spread over their faces.

Many years after this occasion, he still grew irate when telling of his distress in Briggs' office.

"He had received the darned reports O.K.—from the early Frisch-Peierls memorandum onwards! There was the final historical M.A.U.D. report, too, with all the others—locked up in that man Briggs' safe ... because, he said, they were marked 'Most Secret'! No one had read them. I don't believe that mild-mannered old gentleman had read them himself! Not one of them knew what I was talking about. Here it was September, 1941, nearly three years since we knew about fission. Britain and the Commonwealth fighting for life, depending on America to avert the horror of Hitler being first with the atomic bomb, and all that precious time lost . . ."

Oliphant afterward felt sorry for Briggs, near retirement and "landed with the hottest potato of the war." The situation was not as bad as it seemed then; a secret copy of the M.A.U.D. report had reached America three months earlier; it had come into the hands of Dr. Vannevar Bush, recently appointed by President Roosevelt to head the new U.S. Office of Scientific Research and Development. When Oliphant came on the scene, Bush had already organized a committee, formed by the National Academy of Sciences, to look into the prospects raised by the secret British report.

Knowing nothing of this, Oliphant flew west to enlist the support of a man with one of the finest minds of the age.

Ernest Orlando Lawrence was a kindred spirit; his innate curiosity and capacity to see the heart of a problem matched well with the outspoken Oliphant. They knew each other, and they shared a common concern. Lawrence was already distinguished; he had won the Nobel Prize for Physics when only 38 and had invented the cyclotron, which, even then, was taking science along a new path into Hahn's Transurania.

In his laboratories at Berkeley, radio-chemists had made a leap that betokened a new path to nuclear energy. Lawrence had heard the tales coming back from Americans in Berlin and, like many millions of his countrymen, suffered in spirit for the battering Britain had taken from the Third Reich. He needed little prodding from the worried Oliphant.

Not to be overheard on what they had to say, the two physicists walked across the campus, up a slope between Australian eucalypts, to where the gaunt skeleton of the world's greatest cyclotron—a giant 184 inches—was rising against the California blue. Lawrence had none of the tentative reservations of nonnuclear scientists guiding the President. What Oliphant told him—the certainty of Chadwick, Cockcroft, and George Thomson, that a mere ten kilograms of uranium isotope metal would explode with enormous ferocity—left little doubt in Lawrence. On this day, Lawrence told Oliphant, "The waste of time is criminal. It's utter negligence to waste a single day. We know a uranium team has been formed in Berlin. Heisenberg is in charge, with von Weizäcker and Harteck. And, of course, Hahn is there. Not novices, Mark, they're among the world's best! If we know about 235, and if we've found a new element that will explode, why shouldn't they?"

A year before, Rutherford's son-in-law, Ralph H. Fowler, had passed a letter to Lawrence concerning British ideas on trans-uranic elements, suggesting investigation since the British scientists had their hands full. Now, the new facts of fission measurements and the 235 probability crowded in on what had been done in Lawrence's own laboratories.

He took Oliphant to meet some of the radiochemists. They spoke briefly with Seaborg; then Lawrence telephoned Vannevar Bush and Bush's deputy, the president of Harvard, James B. Conant, to make appointments for Oliphant.

Oliphant dined with Conant; it was an affable meeting, with Conant showing interest in all Oliphant had to say. Conant already knew, as did Bush, the details in the M.A.U.D. report, but said nothing to Oliphant. Bush and Oliphant met in New York for some twenty minutes. Like his deputy, Bush gave no hint of the secret knowledge; he played his cards close to his chest, not sure of Oliphant's standing as an official emissary. He did not even reveal that he had offered to engage in a joint uranium project with Britain.

Oliphant flew back to England, unsure of what advance he had made. With the men of the Uranium Committee, with Lawrence, he had stirred a sense of urgency; but with Conant and Bush? Was he just another intrusive foreigner, forgotten as soon as he left?

Four years after Oliphant's meetings with Conant and Bush, the Hungarian-born physicist, Leo Szilard—then famous for his role in the American uranium genesis—gave evidence to a congressional committee: "If Congress knew the true history of the atomic energy project, I've no doubt it would create a special medal to be given to meddling foreigners for distinguished services—and Dr. Oliphant would be the first to receive one. . . ."

In that evidence, Szilard made it plain others would deserve such a medal: ". . . the three men in England who were instrumental in reaching the conclusion that atomic bombs could be made came out of Germany in 1933 . . . refugees who had more time and leisure to think about such remote possibilities as atomic bombs. . . ."

These three men were Otto Frisch, Rudolf Peierls, and Franz

(later Francis) Simon. Szilard said they had been free from the "compartment thinking" predominant in America and in Germany: "It was sheer luck that the Germans did not put two and two together; and we over here did not put two and two together, because the twos were in different compartments. . . ."

All that was still to emerge as Oliphant flew back to Britain, asking himself—what have they been doing over there?

FOURTH PHASE
A Most
Dangerous Daughter

• 1 •

The center of uranium interest came to New York for the first time in January, 1939. On the first Monday in the month, the new Nobel laureate, Enrico Fermi, brought his family ashore from the liner *Franconia;* he was to take up a post at Columbia University, and was to learn, within days, of the revelation he had missed in 1934. Two weeks later Bohr arrived in New York, and it was the day after, January 17, that Leon Rosenfeld unguardedly related the Frisch-Meitner uranium elucidation.

It took a week for Fermi in the Michael Pupin physics building on Morningside Heights to learn from Dr. Willis Lamb, a 25-year-old instructor—who was to win the Nobel Prize for physics in 1955—that Bohr had just leaked great, wonderful news. John Wheeler of Princeton had set up a meeting at which Bohr explained that the by-products of neutron disintegration, identified as barium and lanthanum, were the result of the division of the uranium nucleus into two parts.

Five years had passed since Fermi, Amaldi, Segré, and the others had been bewildered by the collection of active fragments in the target uranium. He wondered why Bohr had not told him when they met; he was not to hear until later of the promise to Frisch, broken by omission.

It was typical of Niels Bohr that, in New York, on his way to take the train to the Washington conference, he should stop off at Columbia University to give his own explanation to Fermi.

However, Fermi had left Pupin before Bohr arrived and, eager to talk to the Italian and thinking he might be working on the cyclotron in the basement, Bohr made his way there. Instead of Fermi, he found the room occupied by Herbert L. Anderson

168 •

and, in his quiet voice that was almost a whisper, said, "Young man, I'm Niels Bohr. Let me explain something new and exciting about uranium atoms. . . ."

Herbert L. Anderson was 25, working on a thesis on scattering neutrons. He had worked under Professor John Dunning, who had made the Pupin Laboratory a center for neutron study; and Anderson's knowledge of circuits had been valuable in building the cyclotron on which he was working. He listened carefully to Bohr, and this moment was to take him to the testing of the world's first controlled uranium reactor and to a further role in the first explosion of a nuclear critical mass.

Following the beautiful reasoning, the logic flowing from the Frisch-Meitner elucidation and the nature of Bohr's liquid-drop nucleus, Anderson thrilled to see how it fitted with his own studies and the equipment he had been preparing. The implications were staggering; the experiments possible were critically important.

Anderson faced no limiting lack of resources such as had hampered Frisch nearly two weeks before. Columbia, one of America's oldest universities, could provide a wealth of technical riches.

Surrounded by equipment that could prove the possibility Bohr presented, Anderson realized that he should consult Fermi. Fermi had returned to Pupin too late to see Bohr. But when Anderson knocked on his door, he took a quick look at his glowing face, and said, "Sit down, please. I think I know what you have come to tell me—eh? I will explain how I see the uranium atoms break in two. . . ."

To Fermi, the notion of the shattered uranium nucleus was an infant; in America, it had yet to be christened "fission." Frisch's experiment was only 12 days old, and his paper on the first experiment was still on the editor's table in the office of *Nature,* in London. Fermi knew nothing of this. He walked to the blackboard to chalk his thinking as he talked, a characteristic to become familiar to Anderson. This was the starting point for a friendly association that would last 15 years, until Fermi's death. Anderson later observed, "Pure physics seemed to flow through the chalk from his fingers on to the board."

All that Anderson had studied for his thesis was now highly relevant. In his mind he was preempting Fermi, taking the uranium from the bottle on the chemical shelves, hooking in the cyclotron to the ionization chamber, which was linked to the cathode tube, the television-like screen. He told Fermi, "Before we do anything, we should first consult Professor Dunning."

The three of them spent an hour, late in the afternoon, trying to coax the cyclotron to cooperate; some gremlin in the circuits barred their efforts. Then Fermi had to leave to catch the train for the Washington conference; Dunning went to an early meal. Anderson stayed on, and only when it was too late to recall them did he remember he could disregard the cyclotron and use one of the neutron sources kept on the thirteenth floor of the Pupin building. He, also, went to eat, planning to return in the evening to perform the experiment.

In that way Fermi missed seeing uranium fission, for the second time in his career.

The light was dimmed in the basement. Professor Dunning, a bluff Nebraskan, was joined by the dean of physics, Professor George Pegram, and Dr. Eugene Booth. By chance, Dr. Francis Slack, visiting from Vanderbilt University, was also there; and all of them worked with Anderson in the preparation. Anderson set a paraffin barrier between the little neutron gun, of beryllium and radium, and the ionization chamber; inside this he smeared one electrode with a paste of uranium. The other electrode, which would receive the increased flow of power by reason of the fissioning atoms of uranium, was linked into the amplifier system which ran on to the screen of the spectroscope.

When they were ready, Dunning moved the neutron source into position, the particles were slowed through the hydrogen in the paraffin, crossed into the chamber, and smashed into the uranium. In the dark, green lines on the screen leapt into jagged, shifting symmetry, wild zig-zag patterns matching bursts of pulsing energy from the exploding cores in the uranium target. Anderson noted, "This was the first time in America."

It was more than that. The four men in the basement were the first to *see* evidence of the unleashed cosmic force in the uranium nucleus. They were 12 days behind Frisch, but only hours ahead of Joliot-Curie, in Paris, and of confirming experiments in California, Washington, and Johns Hopkins University.

John Dunning at once sent a telegram to Fermi. He made

sure that when the stunning news Bohr brought was broken to the Washington conference, it would be known the proving experiment had been first made at Columbia in his laboratory.

One key player was missing that evening—Leo Szilard. He had no status; his was the role of a guest worker, a refugee allowed to keep his hand and mind in trim.

· 2 ·

Szilard was in bed with a high temperature, but his volatile mind could not rest. The news from Bohr had reached him, and his memory surged back to 1933 and 1934—traffic lights in London, the first glimmer of chain reaction, the meeting with Rutherford, the patent passed to the British Admiralty to keep it from wrong hands—and now his fears were revived.

He left his bed and wrote a letter to Mr. Lewis L. Strauss, former Presidential secretary, banker, man of affairs, gregarious collector of potential genius, future head of a vast atomic energy complex. "I ought to let you know . . ." he began. Unexpected news had come; the department of physics where he had been, with Bohr at Princeton, had been "like a stirred-up ant heap." The development could lead to large-scale production of atomic energy and radioactive elements; ". . . unfortunately, also, perhaps, to atomic bombs." He advised Strauss to watch for a coming report in *Nature* by Frisch and Meitner. The letter made Lewis Strauss the first public figure in America to be alerted to the nuclear future.

Four days later, Szilard talked with Fermi on his return from Washington. Fermi was so revitalized from his talk with Bohr about the chance of secondary neutrons emerging from the neutron-shattered uranium core that he returned to New York before the conference ended.

At Columbia, the Hungarian and the Italian joined forces, as far as they were able. Szilard, deeply contemplative, liked to delegate the tedium of experimentation to others while he thought and organized; Fermi was the visionary with busy hands. In the Pupin, Szilard was on the seventh floor, Fermi on the thirteenth, and, although separated, they worked toward the same end—investigation of the nature of fission and proof of a chain reaction. Some new and stimulating aspect seemed to arise almost every other day, and always they were hampered by lack of official support; Fermi's resources were sparse, Szilard's almost nil: he had to borrow money so he could rent a beryllium-radium source, a mere 200-thousandth of a gram, to make some precise measurements. The man who met the bill was Lewis Strauss.

In the same building, John Dunning decided to hunt for evidence of the Bohr-Wheeler notion that uranium-235 was the key to fission; he made a deal with Alfred Neir, at Minnesota, to obtain a micro-amount of separated 235 for assessment of its fission capacity. Meanwhile, Fermi and Szilard continued observing the action of slow neutrons on natural uranium.

On March 3, Szilard was joined in his work by the American physicist, Walter Zinn, who had worked on neutrons at Columbia for some years. Szilard had talked with Teller, in George Washington University, and with Eugene Wigner, at Princeton—another Hungarian exile who also was becoming very concerned about the rumors from Germany.

Szilard had further discussion with Fermi, and then, he and Zinn made their test of whether ordinary water would slow down neutrons sufficiently to cause fission in natural uranium. As Anderson had done, they linked their experimental tank with the amplifier and spectroscope. Szilard noted, "It took two days to prepare and when everything was ready, all we had to do was to turn a switch, lean back, and watch the television screen. If flashes of light were emitted in the fission of uranium. . . .

"We turned the switch and we saw the flashes. We watched for a while, then we turned off the machine and went home. That night, there was little doubt in my mind the world was headed for grief."

Szilard cabled Strauss that the proposed experiment had been performed: ". . . very large neutron emission found. Estimate chances for reaction now above 50 percent."

To Szilard it meant liberation of uranium's energy on a large scale was "just around the corner." That was not to be.

He suffered anguish in the next months; he envisaged the horror of a world dominated by atomic-armed dictators while, at the same time, he was not truly certain a nuclear weapon could be made. In this dilemma, he strove to convince his fellows against rushing into print with results; data in print were a gift of secrets to the enemies of free men, he argued. In the end, he was to campaign against its use in war—but, in 1939, his main drive was to find ways to awake a sleeping government to the looming danger.

It was Fermi, and not Szilard, who moved first to alert American authority to the atomic danger. Two weeks after the underwater test indicated release of secondary neutrons, the Italian—still struggling for a good command of English—went to Washington armed with a letter from Dean George Pegram which introduced him and gave his background to Admiral Stanford C. Hooper, technical adviser to the Chief of Naval Operations. The Nobel laureate did not get to see these exalted men; instead, he found himself in a decrepit old board-room on Constitution Avenue, lecturing officers of the American Navy's technical division.

If Fermi hoped for some tangible response with backing that would expedite the work he saw had to be done, he was doomed to disappointment. Not that there was any truth in the legend, afterward current among scientists, that his naval audience declared, "That wop is crazy!" Indeed, two officers were very impressed; Captain Garret Schuyler, of the ordnance section, wrote a full "Secret" report on the lecture. But Fermi had solid ground only as far as there was a *possibility* of unleashing uranium's energy. He could say nothing on the fission capacity of separated 235 isotopes, nor had he any thoughts on fast neutrons in explosions. It went on record that Schuyler asked him, "What would be the size of a critical mass? Would it fit into a gun breech?"

Facing unexplored mysteries in fissioning uranium-235, Fermi smiled, "Well, it might just turn out to be the size of a small star."

Another officer, researcher Ross Gunn, also glimpsed a horizon. He wrote a memo to his chief, Admiral Bowen, outlining a prospect of a uranium burner driving the engines of submarines, and the Admiral took it seriously enough to decide to back some work on the separation of 235 from common uranium. He offered the Carnegie Institute the token sum of $1,500, which was declined. This was the first repetition of events experienced in Britain; as with Szilard's chain reaction patent, five years earlier, the uranium potential seemed too far-fetched to lift naval minds out of the trough of tradition.

· 3 ·

Szilard's feverish anxiety pushed America into taking the first step to the atomic bomb. He pondered ceaselessly over all aspects of the uranium problem as the weeks of spring and early summer of 1939 went by, and always his thinking came back—if all indications were proven—to the question of supply of raw materials.

With Eugene Wigner, he sought action. Hitler had the poor ores in the old Bohemian mines at Joachimsthal, and nothing could be done about those—but the unique, fabulously rich mine in the heart of Africa was a prize that should be denied to the Germans.

Wigner knew Albert Einstein was a friend of the Queen Mother of the Belgians; could he not write to Queen Elizabeth and seek her influence on the Union Minière Company?

Szilard reasoned that, if Einstein took that step, protocol demanded the American administration would have to be first notified, and that might not be in the best interests of their objective.

His sense of urgency grew fierce. New data had been discovered. He had spent weeks trying to convince fellow scientists not to rush into print and had not been entirely successful; Joliot-Curie had reported on chain reaction; papers from across the Western world displayed wide interest in the subject; the options were closing down on his prospects of keeping uranium's potential in the right laboratories.

It was Szilard's way to reach for the top. If the administration had to be notified—why not the President? Lewis Strauss, because of past political associations, found the White House out of his orbit, but Szilard knew Dr. Alexander Sachs, an erudite economist, who was said to have the presidential ear. That was the path for the Einstein letter.

It was late July before they could agree on the format of the address to Roosevelt; then Wigner could not make the journey to Peconic Bay, Long Island, where Einstein was spending a few days sailing. Szilard had no driver's license; so he asked Edward Teller to join him and drive the car to find Einstein. It was a journey since told a hundred times, and mostly in differing ways; it ended with Einstein signing the letter for Sachs to deliver personally. And that meant further delay—Sachs could not get to see the President until October 11, 1939, ten weeks after the letter was written.

War had come to Europe more than a month before. The weight of concern was heavy on the President's mind; and here was Sachs with a letter, full of strange terms and inferences, from the most famous scientist on earth, saying how recent work by Fermi and Szilard in America and by Joliot-Curie in France—not a breath of Hahn, or of Frisch and Meitner—had made *chain reaction* in uranium almost certain in the near future, and how ". . . a single bomb of this type carried by boat and exploded in a port might very well destroy the whole port together with some of the surrounding territory." And that ". . . the United States has only poor ores of uranium in moderate quantity; there are some ores in Canada and the former Czechoslovakia, while the most important source is in the Belgian Congo. . . ."

The letter urged a speedup in research, the appointment of a special person to coordinate the work, with powers to deal with the problem of securing a supply of uranium for the United States, and it ended with these words:

"I understand that Germany has actually stopped the sale of uranium from the Czechoslovakian mines she has taken over. That she should have taken such early action might perhaps be understood on the ground that the son of the German Under-Secretary of State, von Weizäcker, is attached to the Kaiser Wilhelm Institute, in Berlin, where some of the American work on uranium is now being repeated."

The stilted phrases were cautious. There were only inferences and possibilities; and it was asking too much for Roosevelt to grasp at once this Wellsian concept of adversaries hurling cosmic thunderbolts at each other. Wise as he was, skilled in political infighting, he was ignorant on basic physics—and the age had arrived when national leaders without such knowledge would be ill-informed. However, Dr. Sachs was complimented on his alertness to the possibilities inherent in the Einstein letter and was assuaged with a presidential response. From there on, a legend grew about this day and the letter:

"What you're after, Alex, is to see the Nazis don't blow us all up," Roosevelt is reputed to have said; and he then called his private secretary, General Edwin ("Pa") Watson, and passed him the Einstein letter, saying, "This needs action."

From there, the tale moves directly to the growth of the great American atomic bomb undertaking; but the notion that Einstein's letter sparked the two-billion-dollar uranium project is pure fallacy. Indeed, in approaching the President directly, the uranium problem—which then needed the widest and broadest attack possible—was shrouded in deadening executive secrecy.

William Laurence, the only professional writer officially assigned to tell the story of the secret atom bomb project, said the impact of the Einstein communication was "not worth mentioning". The tragic truth, he said, was it played no part in the national decision to create atomic weapons and that for all its effect "it might never have been written." For two more agonizing years, Laurence noted, in which Hitler might have turned the captured Shinkolobwe uranium into the weapons that would have won him the war, or destroyed civilization, this letter "gathered dust in some White House pigeonhole."

• 4 •

At the end of a career spanning more than 40 years, Dr. Lyman J. Briggs was suddenly lifted to the role of presidential adviser. His basic training in the previous century had been in agricultural science; now, he was confronted by the terrifying responsibility of telling President Roosevelt whether the nation could build a uranium bomb, and what should be done. He was clearly out of his depth.

Briggs got the task because he was director of the government's National Bureau of Standards and, in that respect, would be obliged to keep the subject under the blanket of secrecy the President imposed. Watson gave him two full members of the Uranium Advisory Committee, two service ordnance experts skilled in guns and explosives—Colonel Keith Adamson, a man of short tolerance, and Commander Gilbert C. Hoover, for the Navy. With Briggs' skills light years from the new physics of uranium, it was vital to broaden the base by adding ex-officio members and consultants. Watson called in another member of the Standards Bureau, Dr. Fred Mohler, and Richard Roberts from the Carnegie Institute. Then, since this was a matter referred to the President through Alexander Sachs, he was obliged to invite that gentleman and some of the refugee scientists whose concern had sent Sachs to the President.

The initial meeting of the President's committee, which Briggs held in his office on October 21, 1939—a Saturday, so not to disturb the flow of the Bureau's normal activities—witnessed the first clash between scientists and Army that was to mark the great atomic bomb project to its end in 1945. Stiff and unfriendly, Colonel Adamson displayed resentment to the three Hungarian exiles who faced him, Leo Szilard, Edward Teller, and Eugene Wigner. Albert Einstein had been invited and had declined.

Szilard stated that his immediate purpose was to obtain a supply of uranium oxide and blocks of graphite—as near pure as possible—to establish an accurate measurement of the emission of neutrons from fissioning uranium atoms; these reproduction ratios were vital to a continuing chain reaction and were to be known as the "K" factor, a factor so important that, later, it was jocularly called the "Great God K." Unless enough neutrons were set free in the system, there could be no release of large-scale energy, no chance of an explosion.

Szilard, also on behalf of Fermi and like Joliot-Curie in Paris, had thought of heavy water as a moderator; but there was no source in the world comparable to that which the French had obtained in Norway; so they had turned to graphite to serve the same purpose. He and Fermi wanted money for graphite blocks and for some tons of uranium.

He met skepticism. Teller joined the discussion in typical forceful manner, scouting the question of a flow of money that would be needed in the months ahead. This opened a discussion on general Federal financing; then, displaying his short fuse in dealing with civilian scientists, Colonel Adamson joined the fray with what official records term "a discourse on the nature of war." He declared, "It takes usually two wars to develop new weapons, and, you have to remember, more importantly, that it is not the weapons that win wars! It is morale, the spirit of the troops—not new arms—that wins. . . ."

Eugene Wigner had a high-pitched, squeaky voice. He interjected, squirming restlessly in his chair, "If spirit and not arms is so important, then perhaps it might be quite safe to slash the Army's annual budget by about 30 percent?"

Adamson stiffened, then snapped, "All right, all right! You'll get your money."

In the event, $6,000 was allocated from service funds to purchase the graphite and—at Sachs' suggestion—to obtain some tons of uranium from the head of the Belgian firm, which had set up a subsidiary company in New York.

The meeting agreed to report chain reaction was a possibility, as yet unproved, of which there should be a thorough investigation. It was almost a month before word came back to Briggs that the report had gotten to Roosevelt; he was told it had been "filed for reference."

Nothing was said, or done, about a major supply of uranium. Sachs told the Briggs committee of Union Minière's access to the world's richest source of the material, but no official approach was made.

By then, 2,000 steel drums had been loaded in Katanga for delivery to Port Richmond on Staten Island; they contained 1,250 tons of rich ore and were clearly marked "Uranium. Product of the Belgian Congo."

They were to stand in the open, from September, 1940, for two years before their role was to be realized by a hard-driving general.

· 5 ·

The world's first wisp of separated uranium-235 reached John Dunning's laboratory in micro-amounts—mere millionths of a gram. The sample had been culled by the skill of Alfred Neir, of Minnesota, with the spectrograph. This instrument made possible Neir's weaning of the lighter uranium-235 atoms from their home among the masking uranium-238 atoms. The separation was done with the uranium in gas form, the lighter isotope reacting slightly differently to a magnetic field of force.

During the raw days of February and March, Dunning and his group pored over this fascinating sample. They made historic measurements and recorded reactions to neutrons—until the day the isotope, rejected by Otto Hahn as "too rare to matter," proved itself to be the key to uranium fission. The theory born of Niels Bohr and John Wheeler, at Princeton, was fact.

Dunning and his colleagues presented their findings in papers in the *Physical Review;* the hunt for liberation of atomic energy started a new road in America.

All through winter and into spring, Szilard, Sachs, and others had grown increasingly irritated with the "snail's pace" taken by Briggs. Word of the 235 confirmation went to Briggs on March 11; it was almost a month before he notified General Watson and added that atomic energy was doubtful unless the isotope 235 could be successfully separated.

Even as Frisch and Peierls, under nightly Luftwaffe attack, worked toward estimating the critical mass for an explosion of 235, so the Columbia group obtained physical evidence that the Bohr-Wheeler theory was soundly based.

Among the most ardent and persistent of men at the center of the American uranium scene was Dr. Sachs. Suspicious of lack of response from the White House, he obtained a copy of Briggs' original report of November, 1939, and found it too academic and cautious to have had any effective impact on Roosevelt. While Szilard again went to see Einstein, with drawings for a graphite-uranium experimental pile, Sachs set about organizing a new, direct appraisal. Beyond their circle, interest in fission was spreading rapidly; in the last week of April it held the limelight at a meeting in Washington of the American Physical Society. Fermi was there, so were Neir, Urey, and Dr. Merle A. Tuve of the Carnegie Institute. It was Tuve's interest that was to trigger the truly dynamic approach to the uranium problem.

At this same time, Briggs called a meeting of the Uranium Advisory Committee at his office; Sachs attended and made clear his dissatisfaction: "This work must be prosecuted far more vigorously. If the government will give the green light and plunge ahead, laboratory difficulties and hesitations will disappear. I am impatient with these careful, conservative offerings. If the government is not disposed to undertake full-scale work on this prospect, then I favor trying to get finance from private sources."

Again delaying his assessment, Briggs reported to the presidential aide, General Watson, that his committee did not care to recommend a full-scale drive for chain reaction until the Fermi-Szilard graphite measurements were completed at Columbia.

This letter was written to the White House on Thursday, May 9, 1940. Some 3,000 miles away in Europe, under cover of the Ardennes woodlands, the powerfully mechanized German army was poised to strike; in hours, they would sweep into France, into Holland—and into Belgium, where the yellow hills of the world's richest uranium stood behind the factory near Antwerp.

· 6 ·

The room was not much bigger than a cell. Tucked under the roof of the concrete block called Gilman Hall, on the campus of the University of California at Berkeley, it had an old iron sink and a little cubbyhole with a slanted ceiling that led through glass doors on to a tiny balcony. In this room, Number 307, man produced—from neutron-seeded uranium—a new element.

It started soon after the momentous news from Bohr. Edwin McMillan thought they had come to a blank wall beyond uranium. It had seemed then that the hopes of reaching "trans-urania" had died.

Glenn Seaborg and Edwin McMillan occupied adjacent rooms at the Faculty Club; through the months of 1939 they saw each other often, sharing meals and always talking of the challenges arising from the fission discovery. McMillan one day told Seaborg of a new experiment he had devised to sort out the various energies of the main components of the explosion of uranium atoms—and how this had revealed a radioactivity unlike the rest.

The story came to Seaborg like chapters in an extended serial; piece by piece, from his patient and tidy work, McMillan stitched together a portrait of a substance with a 2.3-day half-life—its radioactivity was reduced by one half in 56 hours. More significantly, in throwing off a beta particle it did not recoil as did the other fission by-products. This was different. Still, its character was enormously difficult to determine; it existed as a few atoms among a million. The experiment demanded all the patience McMillan possessed and all the accumulated skills that radiochemists could apply; and visiting the laboratories at Berkeley at that time, from Washington, was Philip Abelson. McMillan asked him to join his work in the sunny spring of 1940, when he had come to the conclusion that the 56-hour substance could be a neutron-induced offspring of uranium-238. With Abelson's skills,McMillan identified his new material. It was an elemental discovery in the truest sense. He had found the first transuranium element.

At the time Lyman Briggs was informing the President he could not recommend a full-scale drive for a chain reaction experiment with graphite and uranium, McMillan and Abelson were writing their paper for the *Physical Review,* the report which was to distress Chadwick. In it, they revealed that a new world of elements beyond uranium was possible. They named the first offspring, neptunium, after the planet beyound Uranus; and they left the question hanging of what happened after neptunium decayed—were there other stages, further steps, like those suggested by the British, into "transurania"?

Philip Abelson had a vision about where this field of new elements might lead. In a report to Briggs' uranium committee, he observed, "Obviously, the results of these experiments will have a large bearing in the determination of the value of uranium power. It is possible that the cost of isotope separation will be great. The decision to spend a million dollars on a separation plant may well hinge on the results of these experiments. . . ."

From their neptunium was to emerge another element that would be the central feature of a national *$2 billion* undertaking.

The growing prospect of release of energy from uranium fission itself caused a chain reaction of events. At Columbia, Fermi, Dunning, Urey, Pegram and Merle Tuve conferred on a survey of methods of isotope separation. Sachs was prompted to write again to President Roosevelt, by-passing Briggs, to warn that the Germans might capture control of the rich Congo ores. Tuve also attended a meeting of the American Physical Society and came away to write a report for the head of the private research center, the Carnegie Institute, Dr. Vannevar Bush.

Tuve's memorandum urged that nuclear propulsion for submarines was a more immediate prospect than building a uranium bomb—but, he suggested, the interests of the whole nation would be served if a suitable method of isotope separation could be found.

Dr. Vannevar Bush was a shrewd, spry man with a disarming

smile. He was no more a nuclear physicist than was Briggs, but he was far more adaptable, determined, and not too timid to ask for the huge sums of money needed for experimental development of uranium. Bush was skilled in electrical engineering and applied mathematics; he had a part in World War I in the devices for submarine detection, and among his inventions was one that came to be essential to the automatic dials of telephones. When fission came to America, he was vice-president of the Massachusetts Institute of Technology, a post he vacated to play a crucial role as president of the Carnegie Institute. This position soon led to the chairmanship of the National Advisory Committee for Aeronautics. In a sense, he filled a role similar to that held by Tizard in Britain; but, unlike Sir Henry, he was quick to see potential in uranium fission. He had contacts to use in the campaign he planned to mount, and he had the dynamic approach and the grit to organize and assemble the best brains of the biggest scientific community on earth. He became the true driving force in launching the U.S. uranium project.

Bush consulted with Karl T. Compton—his successor at M.I.T.—and with James B. Conant, president of Harvard University. He then approached President Roosevelt and emerged from that meeting with a charter for a National Defense Research Committee which, with himself at its head, would have access to ample funds.

Bush was now virtual czar of American defense science, able to call for service from scientific groups, public and private. He soon brought the Briggs Uranium Committee under his control. There was one rigid ruling from the President—secrecy was to be preserved to the utmost; thus, no foreign-born scientists could serve on the committees. As in Britain, the immense potential in uranium usurped clear judgment; the nuclear province was, suddenly, entirely American property and would remain so with increasing intensity. The fact that most major uranium discoveries had been made by foreigners, been brought to American shores by foreigners, and that the American effort had been initiated and given impetus by the concern and talent of foreigners cut no ice with a President who had an eye on what might happen in Congress. Thus, in 1940, some of the foremost brains in nuclear physics were barred from inner councils by political decision. Bush reorganized the uranium committee—still under Briggs' chairmanship. He rid it of the two service officers, and added physicists Merle Tuve, Pegram, Urey, Gunn, and others. The exiled physicists served only as consultants.

Through the spring and summer of 1940, Szilard scoured America for a supply of graphite bricks—bricks made from the same carbon used to fill lead pencils. It was not an easy task; the graphite had to be as pure as possible, the bricks as dense as industry could make them; and manufacturers had to be given a credible, if untrue, reason for their purchase.

Slowly, supplies started to come through, and when a ton and a half had accumulated in the Pupin Laboratory, Fermi and Anderson commenced brick-building, laying a foundation for the first of generations of nuclear reactors.

They built "piles" of graphite blocks, four feet square and ten feet high. They cut narrow slots in some bricks, and they set a neutron source in the bottom; and through these slots they poked wires of rhodium—element 45. These were placed at increasing heights above the neutron source and left there, for a minute or so, to become radioactive.

The experiment had a drawback. The half-life of induced radioactivity on the rhodium was little more than 40 seconds, and the particle counters could not be installed in the same room as the neutron source if accurate measurement was to be obtained. The scales and monitors were two doors further down the corridor. It meant precision timing by the two men. The rhodium detector would be placed into a slot while Fermi stood with a stop watch. On the count, Anderson would snatch the rhodium wire, and Fermi would grab it and race down the corridor to the counters. He had seconds to place the metal strip in the counter, close the shield, and then watch the flashing lights of the scaling instruments, tapping his fingers with the clicking Geiger counter, working out the loss of neutrons from the "pile" by absorption in the graphite. Slowly, he was building toward the truth of the "Great God K" concept—that graphite would slow down neutrons but would soak up too few to spoil the reproduction rate needed to sustain a chain reaction.

It was soon clear the Szilard's gamble on graphite as a moderator of neutrons had paid off. More and better graphite was needed to obtain tiny increases in the reproduction figures.

Later, the toil of building the piles was extended. Cannisters of uranium oxide would be inserted in the graphite, members of the Columbia football team would be enlisted for the tough work of sealing the cans of uranium powder and manhandling the endless graphite bricks. But the first step to nuclear power would not be taken for another two years, and then in another place.

· 7 ·

Christmas 1940 was sunny and peaceful in California. At his parent's home, at South Gate, Dr. Glenn Seaborg sat in the quiet garden with their house guest, his fellow instructor at Berkeley, Joseph W. Kennedy; they sipped beer, talked of the tragedy in Europe, and of the work they planned to do in the coming year. America was keeping out of the bitter struggle in which now only Britain and the Commonwealth faced Germany; and these two scientists were still traditional researchers, their thinking geared to attainment of knowledge. They now plotted an incursion into the transuranic elements—how to create, to separate, detect, and indentify.

That was how plutonium was born.

They saw a vastly enthralling academic enterprise ahead. Seaborg recalled, "We were after a few atoms in millions. We had to find what were less than traces; amounts as small as a picogram—a million-millionth of a gram! And, when we found it, we had truly to identify it, and then prove it for everyone's satisfaction."

It was to be proved, although not to everyone's satisfaction.

They had been on the slow trail of this unknown since summer, after McMillan and Abelson identified the first transuranic—neptunium. Seaborg had conjectured on what could lie beyond that nuclear horizon. And then, quite suddenly, Ed McMillan had gone to the Massachusetts Institute of Technology, working on the still secret radar.

Seaborg did not at once trespass on another man's ground. He suggested to his young second-year student, Arthur Wahl, that separation and study of the chemical nature of neptunium was a good subject for his Ph.D. thesis; and when Wahl had perfected a technique for isolating pure samples of neptunium,

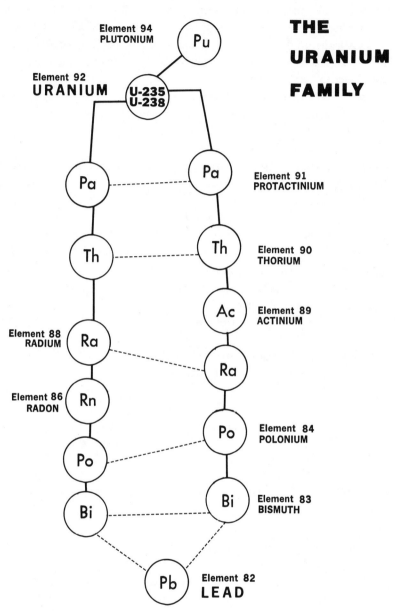

THE
URANIUM
FAMILY

Element 94
PLUTONIUM
Pu

Element 92
URANIUM
U-235
U-238

Pa
Pa · Element 91 PROTACTINIUM

Th
Th · Element 90 THORIUM

Ac · Element 89 ACTINIUM

Element 88 RADIUM Ra
Ra

Element 86 RADON Rn
Po · Element 84 POLONIUM

Po

Bi
Bi · Element 83 BISMUTH

Pb · Element 82 LEAD

Intervening elements are very short-lived

Seaborg wrote to McMillan, ". . . we would be very glad to carry on this work in your absence, as your collaborators." McMillan assented graciously; the ethical path was cleared.

Seaborg remembered his Christmas talk with Kennedy: "Joe and I discussed how we would find our way into the transuranics, theoretically and technically. We decided to try what then seemed a far out idea; we decided to try and produce ponderable amounts of irradiated material to increase the chances of finding these picograms of matter. We thought we could do it using the big cyclotron. Nobody had done that before. Accelerators had been used, but only to get at traces in small samples—and we had access to the biggest cyclotron in the world, the giant machine which the genius of Ernest Lawrence had created. . . ."

Just prior to Christmas they had irradiated uranium nitrate on a small scale. The target material was taken from a bottle on the shelf and plastered on a grooved copper plate; however, this uranium paste was coated with paraffin wax to slow neutrons that would be flung from a beryllium source placed in the path of the particle beam of the cyclotron. This was all easy enough. The great test was to capture the trace amounts of new matter transmuted in the uranium paste.

That was the task set for Art Wahl, under guidance. He had to plan complex steps and follow devious chemical pathways to separate out the confusion of fission by-products, to patiently precipitate the rest of the material until the radiation signals could be read into instruments that would reveal something new—something exciting. And Wahl—working against time to complete his thesis, not even stopping to celebrate Christmas Day—had finally won the chemical fraction of neptunium.

Joe Kennedy's skilled hands built an array of detection equipment; a dozen Geiger counters were fed to amplifiers and linked to an eight-circuit scaler instrument, and there was a device which could count alpha particles and ignore the beta radiation (electrons) streaming from the neptunium.

Early in the new year, the laboratory director, Ernest Lawrence, returned from a visit to the East, where he had aroused interest in the transuranic work. He had talked with Fermi and told him of the guesses Seaborg and Kennedy were making about the elements beyond 93. Lawrence said Fermi had urged that his old colleague in Rome, Emilio Segré, who was in California, should join the hunt. Lawrence said, "Fermi is very eager to help, and he wants to know whether neptunium undergoes fission with slow neutrons. I'd like Segré to help you."

Seaborg, at 29, was a little-known radiochemist; still, he had doubts about the propriety and the security in Lawrence's suggestion. He noted in his diary, "Fermi and Segré are aliens. It is not at all clear that I'm authorised to reveal our results! Still, since Lawrence approves, I assume I can inform them about the new element 93 and of the possibility that, by its emission of particle energy, it could have a daughter substance—which could be element 94. . . ."

It was a helpful union and critical in the outcome. From his sudden bounty of uranium—following the purchase by Sachs from Sengier of 50 tons of uranium—Fermi could be munificent. He sent five kilograms for Seaborg and Segré to work on. Seaborg noted,"Our supplies were always small till then. We could always get half-pound bottles of uranium nitrate from the chemical firms, but few carried anything like kilogram lots. I often had to seek elsewhere for experimental supplies. I once wrote to a man I'd heard had some residues from the old Bohemia mine Marie Curie used; I also chased supplies from Canada."

Seaborg and Segré could now irradiate samples as big as one kilogram: They began leaving this kind of sample in the neutron beam for many hours. They packed the uranium into glass tubes of about an inch diameter and seven inches long and set these inside a ten-inch square paraffin block which they placed close to the beryllium source being struck by the 60-inch cyclotron beam. The neutron saturation was "intense bombardment."

Through the first week of February they pursued the new elements they guessed were in that major sample. They went to work on the Sunday to "milk" the neptunium, and then they tested the residue with Kennedy's new devices, and made clear-cut alpha particle counts. They then discovered alpha radiation was *increasing* in the mother solution of the original irradiated uranium. The signals were suggestive; as yet they had no real proof.

Through the next week more large samples of the Fermi supply were irradiated. All this time Seaborg was worrying over the possible loss of Arthur Wahl; he was a reserve officer, and the Army was about to call him up. The young chemist's precise work had become vital to their team effort, and Seaborg didn't want their big push to be interrupted.

On Sunday evening, February 23, 1941, in Room 307, Gilman Hall, Berkeley, Wahl had the irradiated sample under the final chemical assault that would reveal plutonium.

At the cyclotron, Seaborg and Segre placed the kilogram of uranium under further neutron bombardment that would yield the first sighting of the man-made element. After that, Seaborg went up to the third floor to help Wahl with the time-consuming task of weaning the elusive alpha-emitter from the mother solution of irradiated uranium.

It was exquisite chemical detective work; the long processes of dilution, boiling away oxidizing agents, repeating the processes with the powerful acids, purifying the fraction, precipitating to smaller and still smaller amounts—down to the picogram, to the million-millionth of a gram.

Steam and fumes clouded the air in the small room, and Seaborg stepped through the cubbyhole out to the balcony for a breath of fresh air; he noted it was a new day as he watched lightning split the black sky over San Francisco.

Cooling the final fraction, using thorium as a carrier, Wahl found he could separate the alpha-emitting substance, and, in the early morning, they carried the wisp of matter to their counters in Room 303 and measured the alpha activity.

Seaborg recorded that morning, "With this final separation from thorium, it has been demostrated that our alpha activity can be separated from all known elements. . . ."

In the platinum crucible, the sample they numbered 37—a fragment of matter—fired out some 300 alpha particles each minute. This was different from all known elements! This was no deception, like the fission by-products which had misled Hahn, Fermi, Irène Curie. This was no isotope of thorium, or radium, or actinium. If it *was* an isotope, it was one belonging to the element *beyond* neptunium. Seaborg, elated, said to Wahl, "We've done it, Art! We've nailed down a new element, for sure."

The alpha particles, the low beta count, gave them nuclear fingerprints. There was a new force in being, created by human technology.

Seaborg speculated on this new matter—not so much on what

it looked like, but how would it behave? Was there a true relationship with the elements immediately below uranium? Logic pointed to an element something like radium; certainly, it was as strong an alpha emitter. Could it be as dangerous, as poisonous to living tissue as radium. . . ?

More importantly—how would it behave under neutron attack? Would it fission, like uranium-235? Lawrence said Fermi wanted to know if it would explode under slow neutron attack. Even before they could prove it existed, the speculation had been rife. Lawrence's letter from Ralph Fowler had mentioned that point.

There could be no answer until the work was done with the kind of large sample he had discussed with Joe Kennedy, which he and Segré had positioned in the cyclotron that same night. That should give fractions large enough to tell something more detailed about element 94.

Seaborg later claimed this discovery as "one of the most dramatic in the history of science." He was to argue the attainment in Room 307, Gilman Hall, as the first true realisation of the dream of alchemy, "a unique, synthetic element which, in large-scale transmutation . . . was of overwhelming importance to mankind."

In strict terms, it was one of the deadliest substances known. It was uranium's most dangerous daughter.

· 8 ·

Kenneth Bainbridge flew to England in March, 1941, to be the first American to hear how the uranium bomb could be made and fired.

The young American physicist was an observer, a scientific emissary for his country, one of three on a data exchange mission arranged following the Tizard visit. However, Bainbridge was not, as yet, officially part of the uranium work; his mission was to talk and learn about radar range and direction finding. It was chance he was in London at the time the M.A.U.D. committee held the meeting to discuss the report that would confirm the uranium bomb was feasible. By reason of past membership of the Cavendish, and known to members of the M.A.U.D. group, he was invited to sit in on the uranium review in the Royal Society building.

Firsthand, he heard of the Frisch-Peierls contention that separated uranium-235, hit with fast neutrons, would create an enormous explosion; that the critical mass would be around eight kilograms, which could be obtained from gasified uranium put through a series of diffusion stages.

In the first hour of discussion, Bainbridge knew how far his country lagged behind Britain in this new and critical field of defense study. If there was a race for the bomb, Britain was well in the lead. The theoretical size of the critical mass of uranium-235 was clearly stated; and it was far, far less than the guesses Teller had made in the United States of around 30 tons. Astonished, Bainbridge heard the estimate of about eight kilograms, less than 20 pounds—and how this volume could be reduced by a casing of neutron-reflecting material, which they would call a "tamper." He also heard how Simon was already deep into work at Oxford on a prospective system for the separation of 235 and would have his result by July.

Bainbridge was back at Harvard by May 1, 1941; he cleared his obligations on radar and at once wrote a report on the M.A.U.D. meeting for Vannevar Bush and went to Washington to give an oral report to Briggs.

His message was emphatic. The British were taking the uranium work very seriously indeed; they expected to have a uranium explosive within two years and to solve all the other problems within three years. They were suggesting closer liaison.

Thus, five months before Professor Mark Oliphant flew the Atlantic to find the M.A.U.D. reports still locked in Briggs' safe, Vannevar Bush and his main uranium consultants *knew* details of the British view that the uranium bomb was a probability.

Bush and his associates required further assurance and called for more reports from committee groups. Through the summer, with six months to go to Pearl Harbor, the talk and the debate in America went on with only fragmented studies in Columbia, California, and smaller academic centers.

It was a peculiar procession through the corridors of the chemistry building; swathed in protective clothing, eyes peering from behind screening goggles, hands encased in metal-impregnated gloves, they were forerunners of the world of industrial radioactive processing.

Glenn Seaborg and Emilio Segré, carefully—almost reverently—carried the wooden box containing the leaden beaker with the precious first sample of the new substance—a mere half-millionth of a gram of element 94. Back and forth they went, between their room in Gilman Hall and the nearest centrifuge machine, the high-speed revolving device they used—six times for each sample—to spin the element into a different layer from the materials they had used to wean it from the multitude of fission by-products.

The first weeks of March, 1941, from early morning to midnight, were spent in this complex chemical chase for the elusive radiant fragments. On the first Monday, they had taken out the 1.2 kilogram of uranium from the cyclotron and carried it back to Gilman Hall for the start of the lengthy separation and precipitation processes. They followed carefully planned steps; the instructional notes which still exist show Seaborg's insistence on safety—WEAR GOGGLES and WEAR GLOVES were frequent warnings. The remaining amounts of fission radioactivity were harmful; and Seaborg's suspicion of the poisonous character of the new element increased. Kennedy joined the work and managed the radiation detection apparatus.

Concurrent with the aim of testing the first fraction of the new element for its fission capabilities, Wahl was hard at work in Room 307 winning new samples from freshly irradiated batches of uranium. Seaborg wrote to Briggs and to Abelson, who was then assistant secretary to the Uranium Committee, and, in reporting on the work on the new element, told of the risk of Wahl being called into the Army. He also raised his concern with the University Council, and a letter was sent asking for Wahl's deferment.

Toward the end of the first month of study on the new element, Seaborg received a letter from McMillan that typified the ludicrous security imposed by presidential interest. McMillan told of the visit by Ernest Lawrence, and said, "he is working on obtaining permission for you to talk to Segré about the chemical properties of element 94."

In this odd, official situation, Seaborg and Segré pursued their goal. It was now April, and they had closely followed the behavior of their first sample of the new element.

That sample, some 25 years later, became part of an historical collection at the Smithsonian, in Washington. They fixed it by drying it in a specially built, tiny platinum dish, a mere centimeter and a half wide; the scrap of powder was covered by a thin layer of cement and the sides of the dish cut away. The resulting disk was glued to a strip of cardboard. This was the first sample of element 94—too small to be seen, and once converted from neptunium, it was to be stable for many years; for how long was not then known. When its usefulness was done in 1941, overtaken by purer and larger samples, Seaborg, with the researcher's usual disregard for historical significance, dropped the strip with the sample into an old cigar box. On a laboratory shelf, under the label "Real Smoking Pleasure—Alhambra Casinos," the plutonium sample rested, quite forgotten until Seaborg, when chairman of the Atomic Energy Commission in 1966, remembered and recovered the first plutonium from the tattered old cigar box.

All the work done on the new element at Berkeley, until April, 1941, did not cost the government a cent. Seaborg and Segré, with Wahl and Kennedy, had almost confirmed the fission capability of element 94 before they received a research grant from the Briggs Committee—of one thousand dollars.

Throughout April, Seaborg and Segré pressed ahead with studies of the fission capacity of plutonium, working as fast as the delicate measurements would allow. Indications that element 94 was a new source of energy came early, but since they could not yet be certain how pure their samples were, they could not compare its behavior to uranium-235. New samples of the greatest possible purity were needed, and it was well into May before Kennedy's equipment traced the fingerprints of the new element. They came sharp, clear, and with stunning impact. Seaborg entered his thoughts into his diary: "This now enables us to calculate the slow neutron cross-section of 94 compared with that of uranium-235. . . ."

His estimates had to allow for errors caused by the infinitely small size of the samples and the unknown chemical impurities. Despite the barriers to accuracy, he arrived at a figure that would stand the test of time. He reported to an excited Ernest Lawrence, "Element 94 is almost *twice as fissionable* as uranium-235. The actual figure we have is 1.7 times as great as the fissioning uranium isotope. . . ."

It meant the volume of a critical mass for a plutonium explosion would be less than needed for a uranium bomb. And, importantly, supplies of this new explosive element could be more easily obtained, provided the work on graphite and natural uranium at Columbia showed a sustaining chain reaction was attainable.

Art Wahl worked through the critical weeks of plutonium separation with the prospect of an army call-up hanging over his head. National importance could not yet be ascribed to his work; in effect, he was still a junior member of the university chemical faculty—and that did not impress his draft board.

Not until June 21 was Wahl freed from imminent call-up into the Army Reserve; he was allowed to continue the research that was to give him a place in building the first atomic bomb.

President Roosevelt's edict cloaking the American uranium project in supersecrecy had its effect in the nongovernment laboratories in Berkeley. It could not stop the researchers from discussing their findings, but it did move them to be guarded and to create their own code language.

They talked among themselves of 93 and 94, and called them, respectively, "silver" and "copper"—and ordinary copper became "honest-to-God-copper." Not unil March, 1942, was the decision made on the naming of element 94. There had been full debate before Seaborg reached the final christening. The talk followed the planetary link; after Neptune there was Pluto—but, should it be "plutium" or "plutonium"? Seaborg thought the latter "rolled off the tongue better."

It was settled; the chemical symbol for element 94 was fixed as Pu—and plutonium joined Mendeleev's periodic table.

Seaborg had no doubts that matter could exist beyond uranium: "Our experiments have brought proof that what we have made is chemically different from all known elements. . . ." Seaborg was to see his new element grow to be of "overwhelming importance to mankind . . . to burst on the consciousness of every literate human being. . . ."

From his original million-millionth of a gram, superb technology, driven by the exigencies of war, would produce a tide of plutonium that would expand that first picogram a billion-billion-fold.

· 9 ·

As Seaborg and Segré coaxed the first plutonium data from Wahl's samples, Ernest Lawrence left California and at the Massachusetts Institute of Technology consulted Karl Compton, who shared his concern at the lack of pace and urgency in uranium investigation. His disquiet was increased by Lawrence's excitement over the prospects for element 94.

Compton expressed his apprehension to Bush. America had more physicists and chemists than either Germany or Britain and of just as high a caliber; yet Britain was way ahead, and there was reason to suspect progress on uranium in Germany. Most disturbing, Compton told Bush, was the "tardiness" of the Uranium Committee. He said it met seldom and moved only with "painful deliberation." Too few tasks had been set for the atomic scientists so that this small community was growing restive. This included Urey, who was himself a member of the Briggs' committee. Highly promising lines had been paid little attention, and the uranium problem was so shrouded in secrecy nobody knew what was being achieved in related fields. The chances of achieving an application in the current emergency were impaired, and part of the trouble was Briggs who, said Compton, was conservative by nature and, with responsibilities in other directions, was too slow. Unwittingly, he crossed swords with Bush, for Briggs was now part of the science director's establishment. Compton suggested Lawrence should be made Briggs' deputy for a couple of weeks to see what could be done with a more dynamic drive.

Lawrence went to see Bush at his Washington office and found an angry man behind the brief smile. Bush was under the pressure of well-based criticism; nevertheless, he saw value in Lawrence helping Briggs in the role of a temporary consultant. Within days, Lawrence was able to make some advance.

Within a month a committee had been organized, composed of leading physicists and headed by Arthur Compton (brother of

Karl) of Chicago University; and at Lawrence's insistence the gifted theoretical scientist-teacher, Robert Oppenheimer, was included to present a set of calculations which, unknown to him, confirmed the assessment of a critical mass of uranium-235 that Chadwick had reached in Britain and that was contained in-Bainbridge's report. Compton noted about Oppenheimer's work that it strengthened the committee's urgings for intensified study.

Vannevar Bush added the committee's report to Bainbridge's indication of British leadership in the uranium work; and Bush was a clever administrator, a manipulator of power. By late July, he had persuaded President Roosevelt to create the Office of Scientific Research and Development, with himself as director, with wider power than the NDRC, which now came under his wing. He named James Conant, of Harvard, as chairman of the NDRC, and his deputy.

Under Conant's direction, the Uranium Committee became part of NDRC, as Section One—known as S-1. The presidential decision to create the O.S.R.D. has been rated by some historians as the turning point in the American uranium story. That is not so. The major influence at the time came from the two new developments—the certainty that element 94, plutonium, would fission more violently than uranium-235 under fast neutron assault, and the British decision that 235 could be separated to form a critical mass of around eight kilograms.

Yet the problem of delay was not solved. Two months later, Professor Mark Oliphant arrived in Washington, dismayed and angry at the lack of awareness of the true portent. A nuclear historian, appointed later by the Army to write of the bomb project, made the excuse that American physicists were not used to thinking in terms of science being used for weapons of destruction. It was a dubious concept.

Oliphant's pressure and his urgings in September were followed by a meeting at the White House on October 9, 1941, between the President, the Vice-President, Henry Wallace, and Bush who, despite the British confidence, said the laboratory evidence did not add up to proof, and he could not say an attempt to build an atomic bomb would be successful. Despite this, the indications were strong enough for the two national leaders to order the work expedited; however, Bush was ordered not to take any definite steps with expansion of construction work without further instruction from Roosevelt. Policy decisions, ruled the President, would be made by those three present, plus the NDRC chairman, Conant, Secretary of War, Henry Stimson, and the Army Chief of Staff, General George Marshall.

Twelve days after this meeting, the Compton committee heard Lawrence read a summary of M.A.U.D. work which he had asked Oliphant to write. Lawrence commented on the possibility that uranium might not be a factor in time for the war in Europe: "That is an attitude that is dangerous for us and the free world. It will not be a calamity if, when we get the answers to the uranium problem, they turn out negative, from a military point of view; but, if the answers are fantastically positive and we fail to get them first, the result for our country may well be a tragic disaster."

When the meeting was ended and the form of the report Arthur Compton would write to Bush agreed on, Lawrence reiterated that uranium work in America should be pursued with new urgency,

"We have it from some of the best physicists in the world, in Britain and here, that a uranium-235 explosion is within reach. I believe Seaborg's work will give us a second path to the atomic bomb. We must press on with that work, and with methods of separation of the uranium isotope by gaseous diffusion. We must aim our research at building atomic bombs. . . ."

Speaking to Conant, he was speaking to the third highest power on the uranium work in America—after Roosevelt and Bush. Conant, studious and calm, looked carefully at Lawrence: "Are *you* willing to give the next years of your life to seeing them made?"

Lawrence had a farewell letter from his friend Oliphant in his pocket. One sentence read,

"I feel that in your hands the uranium question will receive prompt and complete consideration; I do hope you are able to do something. . . ."

Lawrence told Conant, "If you tell me that is my job—I'll do it!"

In Lawrence's mind then, was *another* idea that would open a third path to the separation of fission material on a large scale. Oliphant had raised it with him, but he already knew that in a near-vacuum the lighter atoms of uranium in gas might be separated by magnetic fields; and one of his earlier cyclotron models, the 37-inch instrument, could be adapted for a pilot study of this possibility. It wasn't quite certain; there was little, as yet, in this whole uranium program that could be called certain. Extensive work would have to be done; big money would be needed, much more money than the timid Briggs Committe had been ready to demand.

Compton wrote the report for the National Academy of Science, not knowing how his own life would be given over to the same project. The report was lodged with Bush on November 9, 1941, and, in part, it said that "a fission bomb of superlative power would result from the assembly of sufficient uranium-235."

It could not say what was meant by "sufficient"; that was one of the uncertainties to be explored. Compton's report fell on fertile ground. He carried Conant and Bush with him—all three now were thinking in terms of atomic weaponry. The report was also strong enough for presidential approval, and Bush added his suggestions for forming an engineering group and for stepping up the physics research. Production plans were on the move for the first time on a high level; Bush had a rich field of talent from which to recruit.

When he was ready, he called a meeting of the S-1 Committee to reveal the work acceleration and the expanded budget. The date was December 6, 1941; the decision was then announced—American science would work toward atomic bombs.

A few hundred miles north of Hawaii, Admiral Nagumo, aboard the aircraft carrier *Akagi*, was leading his task force south on the mission that, next morning, would detonate into a bitter and bloody war which would culminate in explosions of both uranium and plutonium over two Japanese cities.

The rain of Japanese bombs on Pearl Harbor on December 7, 1941, did far more than destroy a fleet and thrust America into World War II. It shattered a British dream of an Anglo-American atomic police force keeping peace in the world. It demolished all chance of a complete union of British and American nuclear intelligence. It ruined the hopes of a joint project to build atomic bombs because of the cloud of unreality that the very potential of the uranium atom cast over the minds of those with political power. Pearl Harbor gave an urgency to American efforts that quickly outstripped British work on uranium.

In August, 1941, President Roosevelt had made Bush the czar of the American uranium effort. As head of the new and powerful OSRD, Bush was privy to all Britain's atomic secrets through the free despatch to Washington of all the M.A.U.D. minutes.

That month, Bush and Conant conferred in Washington with the chief of the British Central Scientific Office—none other than Charles Galton Darwin, the man who had been on hand in 1911 when Rutherford found the secret of the planetary atom.

The substance of their approach was a golden chance for the embattled British; it was summarized in a letter that Darwin sent posthaste to Lord Hankey in London. The two Americans proposed the building of atomic bombs should be a joint project, government to government. This was clear distinction between a less binding arrangement of cooperative research in both lands. A small joint committee was suggested, with the serious object of seeking how to have uranium weapons ready by 1943. There was to be a stipulation, firm and binding, that neither country could pull out.

Darwin suggested that a three-man British team be sent at once to Washington—the team should be James Chadwick, the M.A.U.D. chairman, Sir George Thomson, and Francis Simon.

According to historical records in London, the reception to this suggestion was "very cool." The British leaders seemed reluctant to enter a firm arrangement from their position of strength and expressed doubts on American sincerity. True, the minister of aircraft production, Colonel Moore-Brabazon, had a vision of joint possibilities. He wrote to Lord Hankey stressing that, apart from conflict, atomic energy could bring worthwhile benefits to mankind. If the bomb was proven, America and Britain, " . . . policing and controlling the world, would have an overwhelming superiority of striking power without need to keep an overwhelming air force." They would know, he said, if other countries were working toward the bomb and would be able to stop them.

Lord Hankey had similar ideas, but, by the end of August, there was blinding evidence that official British attitude was one of doubting American sincerity. Lord Cherwell sent his comments to Winston Churchill: "People working on these problems consider the odds are 10 to 1 on success within two years. I would not bet more than 2 to 1, or even money. But, I am quite clear we must go forward. It would be unforgivable if we let the Germans develop a process ahead of us. . . ."

The odds on uranium had shortened considerably since Tizard's assessment in 1939 of 100,000 to 1. Cherwell then advised Churchill that secrecy favored the construction work being done in Britain or Canada, that whoever possessed the plant making atomic bomb material would dictate to the world. He told the Prime Minister, "However much I may trust my neighbour, I am very much averse to putting myself completely at his mercy, and would, therefore, not press the Americans to undertake this work; I would just continue exchanging information and get into production over here without raising the question of whether they should do it or not. . . ."

On such a patronizing attitude was history changed. Darwin received no response he could pass on to Bush and Conant; not for another seven months would Lord Hankey send a guarded and innocuous reply. By then, it was far too late. Pearl Harbor imparted new strength to Bush's power. Now he could freely tap the vast resources of American wealth, men, materials, engineering skills, and new technology. Once these were harnessed, Britain's lead would be overtaken with a leap; and then would come the day when his clear vision could see, arising from the fission process of uranium and the chain reaction experiments, the need to construct great nuclear powerhouses to produce the weapons materials; and how this would be beyond the hands of academic scientists and their limited resources.

· 10 ·

Coming to the end of his third winter of exile in New York, Edgar Sengier could not understand American officialdom.

For two winters and a summer the huge cache of uranium ores that Sir Henry Tizard and Frédéric Joliot Curie had coveted had sat in the open, unguarded and unwanted, at Port Richmond, on Staten Island. And nobody was interested. Not since Alexander Sachs sought his few tons in early 1940 had there been so much as a glimmer of interest. And now, suddenly, into his office behind Wall Street, on a bleak February day, came this invitation—from the State Department, no less—to attend a meeting in Washington.

They were sponsoring, he read, together with boards of high-sounding company names, a survey of metal resources that might be available, now that America was at war. Not only was the State Department interested; so were the Metals Reserve Corporation, the Board of Economic Warfare, and the Raw Materials Board. For the State Department a Mr. Thomas K. Finletter, special assistant on Economic and International Affairs, was interested in nonferrous metals in the Belgian Congo, and would Mr. Sengier care to submit a report on these possible resources?

Sengier went to Washington early in March. He talked with Finletter and Herbert Feis, also of the State Department, and it was clear they were highly interested in only one metal—cobalt. They asked him if it was possible to double the Belgian Congo's cobalt production. No mention of uranium. However, the 60-year-old mining tycoon was not an easy man to distract: "Gentlemen, listen to me. Union Minière has available material far more valuable and vital than cobalt. It is uranium ores, rich in radium, and bound to be very important to your country in this present war."

The two government officials were polite; they showed a little interest but went on talking about cobalt. Sengier left to return to New York, shaking his head in disbelief.

Early in April, 1942, he again met with Finletter, and, on this occasion, he strongly emphasized the possible strategic value of the ores in the 2,000 steel drums on Staten Island. Finletter showed more interest than before, but only from a security point of view; he knew radium was a highly priced commodity. Incredibly, President Roosevelt's security stricture had excluded the Secretary of State, Mr. Stettinius; and his department was in the dark. Finletter covered his ignorance: "Perhaps, since these ores are so valuable, we might consider having them transferred to Fort Knox, for storage with the national gold reserves."

Later that same month, Sengier wrote to Finletter, with one more try, ". . . as I have told you previously during our conversations, these ores, containing radium and uranium, are very valuable. . . ."

It was a puzzle not to be resolved. Sengier might have felt deluded on the importance of uranium, except for the earnestness of Tizard and Joliot-Curie and Dautry—and the few articles he had read in the American newspapers. A full review of the new source of energy appeared in a *Saturday Evening Post* issue in September, 1940, by William L. Laurence, called "The Atom

Gives Up," which told of the miraculous power science could extract from uranium. And here was the State Department totally disinterested!

Edgar Sengier and his radium-rich ores continued to be ignored by U.S. authorities.

The Eldorado Mining Company's aircraft flew into the pale light of Arctic spring to land the group of mining engineers on Great Bear Lake, near the abandoned mining camp on Echo Bay. They faced hard days and cold nights in Canada's Northwest Territory to bring the flooded pitchblende deposit back into production.

The mine had been a blight on human comfort since its discovery by Gilbert LaBine and E. St. Paul in 1930. Men had worked the underground veins in daunting conditions since 1932, hewing the ores for radium refining at the Port Hope plant, 60 miles west of Toronto on Lake Ontario. This was the development that broke Union Minière's radium monopoly. In 1940 the company closed the mine down.

For two years the workings had stood, flooded and frozen. Now they were to be reopened, and fresh teams of miners would be flown in to extract a few tons of uranium ores to meet the order for 60 tons of oxide for the U.S. Office of Scientific Research and Development.

Vannevar Bush had placed the order with Canada's first uranium mine—even as the frustrated Sengier in New York was making his last attempt to interest the American government in the bonanza of the world's richest ores standing on the wharf at Staten Island.

James Conant called together scientists and engineer members of the expanded S-1 Section, the reorganized uranium committee to meet in his office, on Saturday, May 23, 1942, and the threatening thought that the Germans might well have traveled this same road loomed in their minds and gave their planning a new sense of urgency.

They saw four clear paths to production of nuclear explosives. The zeal and enthusiasm of Ernest Lawrence offered a giant magnetic separation plant for uranium-235; given the go-ahead, he said, such a plant could be producing a hundred grams a day within a year. Gaseous diffusion was also a very strong prospect, and so was a centrifuge process. Then there was the fourth prospect—Seaborg's element 94—plutonium. Compton based an estimate on what Fermi and others had done at Columbia with structures of graphite and uranium oxide, the results of which would be improved when pure uranium metal became available.

Compton's calculations said:

A pilot reactor would be in operation by the end of the year (1942);

Supply of plutonium in gram lots would be extracted by April, 1943;

Kilograms of plutonium would be available by the end of 1943.

It meant, said Compton, bombs could be available some time in 1944.

For the first time they were dealing in tens of millions of dollars: $38 million for a centrifuge isotope separation plant; $25 million for uranium reactors to produce plutonium; $12 million for Lawrence's electromagnetic complex; an initial $2 million for the gaseous diffusion plant. They had, in all, a total of $84 million with annual running costs of $38 million. Compton's report to Bush noted this expenditure would give America "a few bombs by July, 1944, and about twice as many the year after."

But this was at a time when the President was calling on industry to produce tens of thousands of aircraft, tanks, guns and ships, and the priority gradings for essential materials were in the hands of the U.S. Army. Bush therefore recommended to Roosevelt that the Army should be given the job of construction.

Bush wrote his report to the President on June 17, 1942, and, from that day, the Army was to bring about the delays Bush sought to avoid.

On the same day the Bush report went to Roosevelt's office, Winston Churchill boarded a Boeing flying boat at the Scottish port of Stranraer, bound via Washington for a meeting with the President at his family home, Hyde Park, on the Hudson River.

The Presidential aircraft brought the Prime Minister from Washington on the morning of June 19. In blazing heat the pilot made a bad landing, and Churchill climbed down the steps of the aircraft, shaking his massive head, growling, "The roughest, bumpiest landing I've known." He stood solidly, his chin jutting under the homburg, his walking stick held at an angle—like a dress sword. Then he saw the man in the open Ford tourer, grinning and waving a hand in greeting. They were friends in reunion, comrades-in-arms, not so much national war leaders, but a gentleman welcoming a guest to his estate.

They talked in private, but the real business was held over until next day, June 20, when, after lunch, they met in a small room on the ground floor of the family home to discuss the future of the uranium problem. The President sat behind a huge desk, with Harry Hopkins behind him, and Churchill facing the light, sweating in the heat and admiring how cool his two companions appeared to be.

Scientists in Britain had made impressive steps, he said, and were now definitely convinced a uranium bomb could be made for use in this war. "I am here to argue, and to press, that we should pool all our information, work together on equal terms, and share the results—if any—equally between us."

Roosevelt said the groups in America had made advances, too, although no one could say whether any practical result would be won until there were experiments on a wider scale. Neither leader was completely frank with the other. The huge American assembly of brains and money already authorized was not disclosed to Churchill. The Prime Minister knew he was bidding for a partnership from growing weakness. Sir John Anderson, urging a joint project at this late stage, was even then writing, ". . . the pioneer work done in this country [Britain] is a dwindling asset . . . unless we capitalise it quickly, we shall be rapidly outstripped. We now have a real contribution to make to a merger. Soon, we shall have little or none."

Churchill was firm in the decision that the bomb would be built, with or without American partnership. If there was no faith in the project in Washington, then it would be done in Canada where there were stocks of uranium (he did not know that the OSRD had already moved to monopolize these), or, if that country demurred, the bomb would be constructed in Australia or some other Commonwealth country out of the range of enemy bombers.

They agreed, blandly, on the perils in waiting. They knew of

German interest in heavy water—"that eerie, sinister, unnatural term"—which the enemy was obtaining from the factory in occupied Norway. Churchill noted that however skeptical they might be about the scientists' claims, expressed in jargon "incomprehensible to ordinary ears," there was a threat that Hitler could get the atomic bomb before the Allies.

He records in his history of the war that he asked, "What if they should get this bomb before we do? We dare not run the mortal risk of being outstripped in this awful sphere?"

On this threat of common danger, they came to agreement. No written declaration was signed on this highly secret matter, no aide-mémoire was exchanged. They shook hands, agreeing they would join forces to build the bomb—in "common accord," Roosevelt said. Churchill did not dream, Roosevelt did not realize how a military grip was to tighten on the uranium project and bring an all-consuming sense of ownership which, in the end, would deny the "common accord" of the two great war leaders and leave America without a full partner.

Final denial of that "common accord" meant that the great humanitarian nation, the United States of America, would fill the role of the "warring men" Pierre Curie had feared four decades earlier. America became the lone nation to unleash atomic fury on civilians, to carry that stigma into history.

Winston Churchill returned to Britain to face the stunning shock of the surrender of 30,000 Allied troops, with huge supplies, in the North African port of Tobruk. Very soon, he ordered the launching of the first commando attack on the factory in Norway that produced the "eerie, sinister" heavy water.

The Prime Minister was content with the outcome of his mission. The doubts of Sir John Anderson, of bargaining in the atomic arms race from weakness, were swept aside in his mind; he was to claim that it was the progress made by the scientists in Britain, and their confidence that he was able to impart, that had led President Roosevelt to his "grave and fateful decision."

But the most immediate effect to come from the meeting of Roosevelt and Churchill was the clear intent that nuclear war would be launched against German cities—one by one—before the Nazis could do the same to the Allies. Indeed, as late as February, 1945, the ailing President instructed the Army chief running the uranium project, "If atomic bombs are ready before the war in Europe ends—I want them dropped on Germany."

Hitler was the enemy to be assailed first with nuclear warfare; that was the plan. But because of doubts and differences, and further delay caused by the U.S. Army in the critical summer of 1942, Germany escaped atomic Armageddon.

The telephone in Arthur Compton's hideaway cottage in northern Michigan jangled into the heat of a July afternoon. The call was from Berkeley; the voice of Robert Oppenheimer was troubled. Something urgent and disturbing had come up and they had to talk—but not on the telephone. Compton told him, "Catch a plane, Oppie. Get up here as soon as you can. Call me from Chicago and I'll meet the train."

Compton had a lot of time to think while the physicist traveled from California to Chicago: Had the decision been right when, in May, he'd asked Oppenheimer to manage this awesome task of coordinating the theoretical work of bomb design? Yes, he'd known Oppenheimer since he was a student in Göttingen in 1927; and he had felt, when the disjointed work in the different centers was lagging alarmingly, that Oppenheimer might fill the bill. His wisdom, his technical competence, and his charm and persuasiveness with other scientists fitted him for this unifying job. There was some hesitation, though. Compton knew of Oppenheimer's earlier associations with avowed Communists; but before this prospect of special work ever came up Oppenheimer had told him, "The most important thing now is our country's defense. I'm cutting off all my Communist connections. If I don't, the government will find it difficult to use me."

Compton had accepted that, knowing the sheer honesty of the man; he asked him to take on this unique project because of his background, his superb interpretation of mathematical theory—which he had shown at that earlier meeting with Lawrence—and because he could lead men and understand their methods and their attitudes.

When he knew the train Oppenheimer was on, Compton drove the family car to the station. He saw Oppenheimer climb down, brows knitted under the broad-brimmed hat, his eyes troubled. Compton could see he brought disturbing news; still, he had to wait for the portentous details that would confront him with the first of two traumatic decisions that year of 1942—both of which would shape the world beyond the present war.

He drove the car to a lonely beach and, looking out over the waters of Lake Michigan, heard for the first time of the notion that would win the appellation—the "hell bomb."

In the motionless car, Oppenheimer told him of the possibility that had arisen from their delving into the probability of a uranium explosion. The extremely high temperature of the fission release of energy could be used to trigger a fusion reaction in deuterium; and it seemed likely that, pound for pound, the hydrogen would give a much bigger explosion—and would not be restricted in size as was the critical mass in fission. There did not seem much doubt that, if a fission bomb could be assembled, the path led on to hydrogen explosions—thermonuclear weaponry of enormous savagery.

They had gone into it thoroughly—the whole group, including Teller, Fermi, Hans Bethe, Robert Serber, and the others. They had even considered there might be peril to the whole environment in such an explosion—whether it would cause uncontrolled chain reaction and set fire to the earth's atmosphere.

In the torrent of thought that flooded his mind, Arthur Holly Compton wanted to reject the idea as fallacious—but these men included some of the best brains in America. They were not fools. Obviously the chance was there—and it would be there for the enemy. Yet if it could endanger man's habitat, destroy all life, would that be worse than living under Nazi domination of the world?

Oppenheimer needed guidance. No—he needed some firm instruction. Compton told him, "The simplest bomb is your first objective. It has to be attained before the other prospect has any currency. But Oppie, just the same, carry on with the hydrogen fusion calculations. They cannot be ignored."

In this way the superbomb hovered in the backrooms of the atomic laboratories, not to emerge as a prospect until the war was won—and the peace was again threatened.

· 11 ·

As the weeks of July and August, 1942, went by, Vannevar Bush's concern rose. Since his first approach to the Chief of Staff,

General George Marshall, for Army involvement in the bomb construction, his liaison had been with Brigadier-General Wilhelm D. Styer, head of the Army's newly created Service of Supply, which held tight control over priority ratings and material resources. Since then, Bush's S-1 Section, his uranium project, had become embroiled in the vast mass of competing priority claims. Service of Supply, with a staff of one million, was then completing some $600 million worth of projects—each month.

It was not that Styer didn't know the potential of the uranium bomb. Once General Marshall had cleared Styer, Bush took him into his office, briefed him, and then left him with files, taken from the OSRD safe, to read. From an adjoining room, Bush could hear the exclamations—"some of them profane," he recalled—and knew Styer was passing through the stages that afflicted each mind new to uranium's potential. It was a process he was later to observe in President Harry Truman—interest, skepticism bordering on disbelief when the figures were read on energy release, then surprise, and finally unbridled excitement.

Styer knew the stakes. The attitude of the military supply mandarins was expressed in their official history: "True, they knew the atomic bomb might win the war—one day; but, in the meantime, the Army had to make sure the war wasn't lost."

The speedy action Bush and Conant saw as vital was not reflected in the priority ratings; the bomb project came much below the program for synthetic rubber. As well, appointed officers disputed whether four methods should be followed for production of bomb materials—why not only two? Then the selection of the Clinton site in the Tennessee Valley Authority area was held up on the question whether it should be requisitioned before the final research was done.

It was true that Bush's deputy, Conant, had had second thoughts and had been putting off a final, all-out decision. Like officers in the Army, he had questioned whether the atomic bomb would become a viable wartime proposition, but that was only until he had heard Oliphant and Lawrence. Conant had said to Lawrence, "I have argued that the atomic bomb should go into wraps for the duration, but you come along with a definitive plan—and now we must do it first."

He had then said to Arthur Compton, when writing the National Academy of Sciences report, "Get your report into Bush's hands as soon as possible—we musn't lose a day!"

Added to this, Bush had high respect for the quality of German science. All those long summer days the thought of what they might be doing in Berlin was with him—and the months slipped past. In September, his S-1 members were near-desperate for action, and Bush's impatience was growing into anger. Then came his onslaught; letters calling for urgency—to the President's office, to General George Marshall, to Henry L. Stimpson, Secretary of War—to Styer, to the War Production Board, protesting at the lack of action by the Army, which had wasted months in delaying acquisition of critical sites.

Styer came to his office in mid-September, and said, "I think I have the right man for this job." Involved as he was in his complex daily struggle to advance national scientific defense structures, Bush paid little heed.

The quartz filament was less than five inches long and hair-thin. Under the gravity pull of the trace of yellow-green powder, the glassy fiber bent slightly. This tiny force was measured, calibrated with the aid of a microscope, as 2.77 millionths of a gram. It was the smallest measure of weight achieved in America; it was the first weighing of plutonium.

On September 10, 1942, Glenn Seaborg and four of his plutonium group watched this unique event in Room 405 on the fourth floor of the Jones Laboratory, in the University of Chicago. The known gravity of this wisp of the first man-made element was to be the touchstone to the evolution of its chemistry, to the revelation of its physical behavior, to the data on which future plutonium work would be based.

On April 19, 1942, a cold Sunday morning, Glenn Seaborg had arrived in Chicago with Isadore Perlman to plan the chemical steps they would take. They were now integral with Bush's OSRD; in Le Conte Hall Army officers had instructed them on security restrictions. In Chicago, with Seaborg heading the plutonium group in Arthur Compton's so-called "Met-Lab" (metallurgical laboratory), secrecy would be rigidly observed; no more would anybody be able to wander freely from one laboratory to another, without question.

Seaborg was just 30; he was to spearhead a thrust for knowledge of a unique substance—to follow the steps of separation of element 94 from irradiated uranium, the pursuit of its potential, and its preparation as a metal to be used in war.

There was a growing number of top-flight chemists assembling under Compton's wing; many of them were drawn to Room 405 when, four months after reaching Chicago, Seaborg could display the first sample of plutonium to be seen on earth. Seaborg wrote of August 20, 1942, with customary restraint, "The most exciting and thrilling day I have experienced since coming to the Met-Lab. It is the first time that element 94—or any other synthetic element, for that matter—had been exposed for the eye of man to behold ... my feelings are akin to a new father engrossed in the development of his offspring since conception. Counting from the time that uranium oxide was first bombarded with neutrons on December 14, 1940, to produce this 94 isotope, the gestation period has been 20 months. . . ."

That was an "occasion," almost a holiday. Some brought cameras, and shutters clicked under floodlights; there was merriment and congratulations. And behind it were Seaborg's nagging worries: "I don't think it wise to expose any one person to too much radiation."

There were accidents that summer. The plutonium was almost lost—twice. Seaborg's assistant, Louis B. Werner had placed the glass tube in a centrifuge—in the main part of the Jones Laboratory—and was back in the micro-chemical laboratory of Room 405 when he "heard the clatter." Something had come apart. When he reached the machine and stopped it, the precious liquid containing the plutonium was dripping through the motor bearings, on to the floor. Werner reported to Seaborg, "Fortunately, by sopping the liquid up with towels and sponges, we could digest them in acid and recover almost all the plutonium."

In fact, recovery of the element was priceless experience for plutonium extraction when it came in larger scale from the first big reactors the Army was to build in Clinton, Tennessee.

Isadore Perlman—who had been an undergraduate with Seaborg, and was later to be a leading nuclear scientist in Israel—knew a similar moment of panic. He had worked on initial steps of separation down to where the micro-chemistry took over, and he had a beaker half-full of liquid containing the plutonium. He put it away in a cupboard for the night; somehow, a lead brick tipped over, and next morning he "found the beaker broken and the precious material spilled." Again they were fortunate—the beaker had been standing on a Sunday edition of the *Chicago Tribune*. Perlman joked, grimly, "A half-liter of anything is nothing for that paper to absorb!"

He got the largest evaporating dish he could find, and in that he kept a "witches' brew" going with nitric acid for days on end. It caused delay, but the plutonium was recovered. Later, he said, "I remember vividly how the print floated. I was very grateful for that newspaper. ... I am probably the only person to *digest* a Sunday *Chicago Tribune* so thoroughly."

These events were landmarks in Seaborg's mission to explore and elucidate the explosive substitute for uranium-235. There were still some chemists who argued that this was not a distinct element but was probably another isotope of uranium. And there was further doubt. He was to write during that climactic time of chemical searching: "Not everyone shares Fermi's confidence that his [graphite-uranium] pile will chain react and produce element 94. And without a working pile, we will never be able to produce 94 in much greater yields than we get from the cyclotron—and it would probably remain a novelty for years to come. Without the pile, the dream of atomic plants will come to naught ... and without 94, the only possibility of producing a bomb will be to use the uranium isotope."

Seaborg and Werner agreed the first wisp could not be "tied up just for exhibition;" that would cause further delay. They decided on "harmless deception" and made up a simulated—and more impressive-looking—solution of 'plutonium' from green ink; this was prepared in a small glass tube each day before some important dignitary was due. Strangely, the green ink overnight turned a slight plum-purple—strange because it afterward emerged that plutonium, in a stage of oxidation, was actually purple in color.

Seaborg explained, "True, we fibbed a bit in using green ink for visitors, and we even used aluminium hydroxide dyed green for the more informed people; but I always softened my conscience by saying to them, 'This represents a sample of plutonium hydroxide,' and the significance of that was never noticed."

In the years to come, as the plutonium samples grew bulkier, from supplies of uranium irradiated in the massive reactors then being planned, thoughts of deception would be forgotten. Time would show the substance to be so dangerous that remotely controlled complexes, with underground caverns, would be needed to avoid hazards to human lungs, kidneys, and intestines, and would compel measures and intensive health checks such as no human activity had hitherto demanded.

All that was ahead in the massive construction drive that was to bear the name Manhattan Engineer District—later, the Manhattan Project.

· 12 ·

Tact was not a prominent weapon in the armament of the incredible Leslie Richard Groves. On September 17, 1942, Colonel Leslie R. Groves of the Corps of Engineers, at 46 was one of the chief construction bosses of the Pentagon, a soldier long in the supply service who was looking forward to being sent overseas for the invasion of North Africa. Now, the hope was dashed; for that morning General Brehon Somervell told him he had been assigned to a task that meant he had to stay in Washington and added, "If you do this job right, it will win the war."

From the applications he had seen on priorities for the OSRD uranium project, Groves knew at once what the job was; disappointed and disgruntled, his spirits sagged, and he remembered saying, "Oh! *That* thing."

The blow was softened by General George Marshall, who ordered Groves to be promoted to brigadier general and to "take complete charge of the atomic project."

Back in his office, Groves met and talked with Colonel Kenneth Nichols, who had been attached to OSRD since the earliest days of Army involvement in uranium. He asked Nichols: To what extent was the project based on "plausible theory ... unproven dreams of the scientists?" And what was the situation for raw materials, for the ores that would be needed?

Nichols' reply "horrified" Groves; there were mountains of theory, little proven knowledge. Groves assessed the prospect of the atomic bomb as founded only on "possibility rather than probability." The engineering concepts were "unprecedented," the raw material position "uncertain." He decided to go direct and unannounced, that afternoon, to call on Vannevar Bush. It was a chilling encounter. Groves, claiming now to be "in complete charge," demanded answers to secrets under the presidential shroud. At once, there was a confrontation.

The official record of this meeting shows Bush under tension, "mustering all his self-control" in the face of the brusque, bulldozing approach of Groves. And Bush recalled the words he had written days before to the White House: "Faced as I am with the unanimous opinion of a group of men I consider to be among the greatest scientists in the world, joined by highly competent engineers, I am prepared to recommend that nothing should stand in the way of putting this whole affair through to conclusion, on a reasonable scale, but at maximum speed possible, even if it does cause moderate interference with other war efforts."

If he was right—and he was—Groves would be the type of commander who would want to put the physicists and chemists into uniform. What collision course could be worse for the now-lagging uranium program?

Bush saw the general, red-faced, leave his office growling about a "most unsatisfactory meeting," and he immediately telephoned General Styer to voice his apprehension. Then he wrote to Harvey Bundy, liaison for the Secretary of War, Henry Stimson, and said, ". . . having seen General Groves briefly, I doubt whether he has the tact for this job. . . . Styer agrees with me that the man is blunt and aggressive, but that his other qualities would overbalance this. . . . I fear we are in the soup!"

Further enquiry added to concern; Groves was a contentious spirit with a sharp tongue and brutal humor which often offended fellow officers. It was a bad beginning—it was to be their worst encounter.

That same day of their meeting, the bluff soldier's intelligence and dynamic approach, those qualities that *did* "overbalance" an unfortunate manner, showed themselves in action; General Groves took steps to ensure that America would have enough uranium for its atomic bomb project.

He learned that Edgar Sengier was asking the Canadian Eldorado Mining Company to refine 500 tons of ore into high-quality uranium oxide. Within the hour, his aide, Colonel Nichols, was on the way to negotiate with the Belgian, in New York. Orders preventing movement of uranium ores out of the United States were imposed. Proposals to have the Canadian government purchase a majority of stock in the Eldorado Company were confirmed; this ensured purchase of ores and refining at the Port Hope plant would be an intergovernment arrangement. Almost as speedily, Groves moved to acquire the disputed Tennessee site.

Within two days, top priorities were attained for the uranium work, and a bottomless coffer of money was opened for what was to become the greatest organization of science for war, in history. Differences of attitude were also quickly settled with Vannevar Bush.

Colonel Nichols met with Sengier on September 18 in his office at the African Metals Corporation in New York.

By mid-morning they reached agreement, Nichols writing the text of their negotiations on a yellow notepad. With relief and delight he watched Sengier initial each page of a document that would ensure the raw material for the Army atomic bomb project. There had been doubt, at first.

When Colonel Nichols arrived at Sengier's office, Sengier's greeting had been quite abrupt: "Are you here to chat, like the others? Or are you here to do business?"

Through the coming years of the Manhattan Project, Nichols would be noted for diplomacy, and tact: "Business, of course, Mr. Sengier! I'm here to buy your uranium, not to just talk about it."

Sengier, much relieved, then told Nichols how, through the months, department officers had treated his approaches on the ores' value with bland indifference. And this offhand attitude had made Sengier suspect the Shinkolobwe ores were being held in New York without any real commitment to buy. Neither of them knew the State Department had been kept in ignorance of the uranium problem by presidential order.

Sengier soon made clear to Nichols he knew the basis for this sudden government and Army interest. In Britain and in Europe, he had been alerted that one day these ores would be crucial; and this was the day. Free of security precautions, Nichols could negotiate from the strength of ensuring an Allied victory, of liberation for Sengier's homeland from the Germans. There was no haggling over price; they agreed the ores would be paid for at prevailing prices, with full assurances that Union Minière would be treated fairly and Belgium would share, after the war ended, in any benefits stemming from the atomic projects.

A smiling Nichols left Sengier with a handshake and the

signed notepad agreement in his briefcase. In an hour, he had completed what governments had failed to do; he had provided for wartime needs for the coming uranium complexes—and more. He had the signed promise from Sengier that surface uranium ores, hand-picked by the Congo tribesmen and still in Katanga, would be shipped by rail across Africa to Lobito and then by fast freighters to New York. That promise meant the United States was to acquire the richest stock of uranium on earth. Two cargoes were lost on the Atlantic crossing from U-boat attacks, but, in all, America gained more than 4,000 tons of the high-grade ores, more than four-fifths of all uranium used in the Manhattan Project. Yet the Shinkolobwe mine played a far more dominant role than these figures indicate. No other uranium, before or since, matched the energy value. From each 1,000 tons of the yellow-green rocks, the Port Hope refinery, on Lake Ontario, processed 680 tons of high-quality oxide for conversion to gas in the diffusion separation plants built for the extraction of uranium-235 and for the metal slabs and canned slugs to fission in the great uranium reactors the Army was to build to breed plutonium.

The average oxide yield of 68 percent compared to less than one percent from the Canadian Eldorado workings and to much lower percentages from other sources then available.

In order to cover this massive and critical purchase, General Groves had to arrange a secret fund through federal channels. A deposit of $37 million was laid against his name to be drawn on, without question by bank or Treasury officials. Sengier also had to be given special security arrangements for bank payments made to Union Minière, the amounts of which have never been disclosed.

The deal in Sengier's office on September 18, 1942, was followed by immediate action. Nichols had the ores moved at once to the Seneca Army Ordnance Depot in Orange, New Jersey; from there, the uranium went in 100-ton lots to the Canadian Port Hope refinery to become high-grade oxide. From this was made the few dozen pounds of metals that helped bring a sudden end to World War II.

· 13 ·

The site of the first uranium reactor was marked out with a stick of chalk.

It was a circle, drawn on a sheet of rubber-coated balloon cloth laid over a wooden frame. Within the area of that 19-foot diameter chalk circle was erected, layer by layer, in less than a month, the structure of graphite bricks and uranium that first put nuclear power into human hands.

The historic happening took place in a squash court on the campus of the University of Chicago. So certain were the physicists that nothing would go wrong, only the normal brick walls and the balloon cloth stood between the crowded southern suburbs of one of America's biggest cities and the fissioning 50 tons of oxide and 6 tons of uranium metal, set in 350 tons of neutron-permeated graphite slabs.

The first reactor should have been assembled in a building some 23 miles away, in the Argonne Forest Reserve, but union disputes halted construction.

Three years of hampered exploration had led to this new impasse. At Columbia, and elsewhere, Fermi, Szilard, Zinn, Anderson, and others had persisted with the countless experiments, building mock graphite assemblies, testing neutron levels, working out controls for the atomic fire they were to light. Always seeking higher and higher purity of materials, working to eliminate substances which soak up precious neutrons, they assailed the "Great God K"—searching for the combination that would give levels of neutron reproduction to allow a controlled chain reaction.

Earlier that year, to further coordinate the national push into the uranium project, Fermi and Szilard and their associates had moved to the Met-Lab setup under the guidance of Arthur Compton at Chicago—part of the overall OSRD program, but without the power of decision.

In October, 1942, with the first Shinkolobwe ores in the form of oxide and uranium metal, Fermi had enough material to attempt a controlled fission reaction. Eight years had gone by since he first bombarded uranium; now when there was so much at stake, a union squabble, barred this step toward scientific immortality. It was quite insupportable. He went to Compton with his plea: "I believe we can safely build this pile, right here in the university."

It was a terrible decision for Compton to face. No man had been known to attempt this before. Within the fearsome binding forces of the uranium nucleus there were still unsolved mysteries. Compton recounted his awful dilemma in later times: "Under the conditions of release of nuclear energy of such vastly greater power than anyone had previously handled—and in the midst of a great city—some new and unforseen development might appear."

Within the bounds of his nuclear training, he could not see how a "true explosion could take place, like that of an atomic bomb," not in a pile of moderating graphite. Yet there was something he saw as more daunting: "The amount of potentially radioactive material in the pile would be enormous; and anything that would cause excessive ionizing radiation in such a location, which might greatly affect the city, would be intolerable."

Compton well knew he could go to higher authority to settle the dilemma. But that would have been unfair, even cruel, to university President Hutchins, who was in no position to weigh the peril and the hazard; and, faced with that decision, Hutchins could only serve the university's well-being—he would have to say "No." And Compton felt that answer would be wrong; against it was Fermi's confidence. Compton said to him, "Let me hear your analysis."

There was the estimated size to recount, controls, instrumentation, fail-safe methods of neutron-dousing material in liquid form to pour over the assembly should the fission of the uranium get away. Compton said, "I must assume the responsibility. Go ahead with your experiment. Make it in the squash court under the West Stands."

In mid-November, there was a meeting in Washington of the S-1 executive to review the OSRD program. Compton held his peace until the discussion had gone on for several hours; then he announced that building the first chain reacting uranium pile had been started—in the university's squash courts.

Compton knew it would shock the cautious Conant; he saw his face whiten. General Groves hurried from the room to telephone for verification that the Argonne building would not be available that year. Their view is in the official record: "It was a gamble with a possibly catastrophic experiment in one of the most densely populated areas of the nation."

It was, however, to be a triumph for the theoretical and empirical researchers. Doubts of Compton's judgment and reliability were forgotten sixteen days later.

There were two crews working a 12-hour day, seven days a week. Herbert Anderson—who had spent weeks finding timber for the frame and the special balloon cloth to encase the reactor and shut out air (if that became necessary)—led one group; Walter Zinn, the other. They planned to build the pile of graphite and uranium in as near a spherical shape as possible. With Fermi calculating the most effective positions for the uranium, they laid the first bricks on November 16. More than 40,000 graphite blocks had been drilled to contain half-spheres of uranium oxide, crushed into temporary solid form with a mounted press. The toil was arduous, dirty, and too much for the scientists and technicians alone; so a team of tough Chicago youths was enlisted, and that brought trouble. The young men had finished high school and were waiting for the Army draft; they were interested in the money and little else. Some of them were diametrically opposed to a fair day's work for a fair day's pay.

Physicist Albert Wattenberg recalled, "The kids created real challenges. Some had a negative commitment to work and were very enterprising."

They tightened up bearings on the lathe so that it jammed; they found ways of ruining the pulley belts, of breaking the press, of stealing graphite bricks, of causing chaos in the whole progress. One of the worst offenders was detected, but he was drafted by the Army before he could be fired.

There were deeper problems; when the first boxes of inch-cube uranium metal arrived and were opened, the oxygen on the metal's porous surface started to smoke and burn. Dry ice was poured over the metal, but not before some had been ruined for their experiment. There was also the cold to contend with. The snow was deep outside, the temperature fell to 30 degrees below zero—and the squash courts had no form of heating. Zinn found some gas fires, shaped like logs, but these left fumes in the air; so, they punched holes in oil drums and lit charcoal fires, and the smoke filled the place and made breathing difficult.

Walter Zinn recalled, "Scientists and technicians used physical activity . . . but the security guards had to stand in one place, barring the entrances to our secret work. The university came to the rescue. Big league football had been banned from the campus some years before. We found a locker full of discarded racoon fur coats . . . we had the best-dressed collegiate-style guards in the business. . . ."

Fermi, Anderson, Zinn, and others made their daily calculations of how big the pile would have to be before it could go to the point of being critical, of reaching a self-sustaining chain reaction. It was built with the cadmium rods in place to soak up excess neutrons while they added layer after layer of bricks and uranium. Cadmium atoms are not made unstable by the intake of excess neutrons.

On the night of December 1, Anderson halted his shift; the fifty-seventh layer of bricks and oxide had been laid. Earlier, he had agreed with Fermi not to go beyond this point. The top of the pile was close to the ceiling, but that was not the reason for ceasing the work of building the pile. Fermi had said to Anderson, "Promise me that when you have laid that section you will make your measurements—and then go to bed."

Anderson experienced the temptation Fermi had foreseen. With the fifty-seventh layer completed, with only one cadmium rod left inserted to soak up the neutrons, he knew, "It was a great temptation . . . to pull the cadmium rod, to be the first man to make a pile go critical!"

On the morning of December 2, 1942, there were many calibrations still to be made, but there was true excitement. And this new "creature" was to show a surprising face; it proved to be the world's biggest barometer. Just to open a window or a door would affect its activity—with colder air, it moved into faster neutron gear. Temperature was thus seen to have relationship to fissioning uranium.

The last cadmium control rod had been designed and built by Wally Zinn; they christened it "Zip." It was operated by a rope on a pulley and was weighted inside the pile so that in an instant it could be dropped into place to dampen the flow of neutrons.

Through the hours, Fermi stood at the center of a group of more than 40 scientists. With a slide rule, he checked each reading of the Geiger counters for the increasing levels of radioactivity as each control rod was edged outward from the heart of the uranium fire.

After lunch, Compton stood with the group on the balcony, overlooking the black graphite mass. Slowly, the last control rod was edged outward. The clicking of the counters increased to a rattle, then to a roar. Fermi gave the signal to switch to a chart recorder with a stylus moving upward to display the mounting intensity of the neutron storm, now far too intense for the mechanical counters to register.

Fermi was bent over his slide rule; with the counters silent, Herbert Anderson remembered, "It was awesome ... we were in the high-intensity regime, the neutron count increasing more and more rapidly."

Then Fermi raised his head, held up a hand: "The pile has gone critical."

Still he waited. Many shifted restlessly; why didn't he order the control rod be put back, to shut it down now? Fermi was motionless, calm, watching the seconds and the graph stylus climbing higher on the chart. Another minute; then another. The clock stood at 25 minutes past three in the afternoon. Fermi called to Zinn, "Zip in!"

The cadmium rod fell into the reactor, absorbing neutrons; the graph stylus dropped down the chart. The first atomic reactor had obeyed the hand of its creators.

Compton shook Fermi's hand. Fermi grinned at Szilard. Eugene Wigner produced a bottle of Chianti and paper cups. They sipped the wine in a toast and then all signed the label on the bottle.

Compton went out into the raw cold of the winter afternoon to walk across the field to his office in Eckhart Hall, his thoughts fastened on the future. Here was a new source of power: a golden, glorious age was dawning, with cheap, limitless energy, for homes, for industry, for the less fortunate peoples of the world. His gamble had come off.

He called Conant at his Washington office, devising a code in his mind to preserve secrecy.

"Jim, you'll be interested to know—the Italian Navigator has just landed in the New World!"

"What? Already?"

"Yes, the earth was smaller than he thought. He arrived earlier than expected."

"Were the natives friendly?"

"Yes, indeed. Everyone has landed safely—and happily."

· 14 ·

Dr. Glenn Seaborg met one of the witnesses to the first chain reacting pile operation in a corridor of Eckhart Hall. His eyes told the tale. Fermi's pile was a success. The path was open to the wholesale creation of plutonium; Groves would now drive for the construction of the immense reactors to breed the transuranic element. Immense, indeed: Fermi's reactor, even if it ran for 10,000 years, would not produce enough element 94 for one single bomb.

Seaborg knew he would have to lead his team into devising means for industrial production of plutonium of the highest purity. (Even so, as extensive as their skill became, the matter of purity—rather the lack of it—was to prevent the fission-firing of plutonium by the method that would be used in uranium-235.)

A further thought went into his diary that night: "We have no way of knowing if this is the first chain reaction achieved. The Germans may have beaten us! I wonder—are they aware that elements made in the chain reaction of uranium isotopes can be used in an atomic bomb? And, if they do have a pile that reacts, will they use it to generate power, or produce vast amounts of radioactivity as a military weapon. . . ?"

The power of the shattered uranium atoms was to plague Seaborg's life for years. Exquisitely accurate chemistry was vital to the separation of one part in 100 million to extract pure plutonium from the neutron-bombarded uranium. For the Manhattan monoliths, he would devise means of extraction of fission by-products, every ton of which would exude 10 million Curies of gamma radiation—instantly lethal to human beings.

Where the Columbia River flows around a wide stretch of arid land near Pasco, in southeast Washington State, Manhattan Project officers claimed dozens of square miles to build the huge successors to Fermi's Chicago pile. In the underground caverns, behind thick lead shields and concrete walls, plutonium would be extracted by Seaborg's formulae. He would learn that his marvelous discovery was among the most dangerous materials known. At this sprawling plant, to be called Hanford, modern alchemy would reach new peaks in what would be named by scientists, Devil's Canyon.

Unparalleled and complex industrial innovation would be created to deal with this substance which he later described as "One of the most deadly substances known, it has unusual—and unreal—properties."

A whole year after the Fermi pile went critical in Chicago, Seaborg began to receive samples from the first of the big uranium reactors. He walked into the plutonium laboratory on the fourth floor of the Jones building in Chicago toward the end of 1943, and it struck him that men were casually handling deadly danger. He told this writer, "We had hitherto been working with millionths of a gram. Suddenly, we had samples a thousand-fold bigger. We had milligrams. And I saw the fellows working there with open beakers and I called a halt. I demanded that we have proper protection. I knew this element was a strong alpha-emitter, just like radium, and I believed it to be dangerous, like radium—and I knew that from then on we'd better be careful. . . ."

Within the next year the peril was well known. An official report by the Manhattan staff stated, "The metabolism of plutonium is similar to radium. . . it is deposited in the bones where its alpha radiation may cause bone sarcoma. . . . Among other body organs, the heaviest deposition occurs in the kidneys where its radiation, in sufficient quantity, causes destruction of . . . kidney tissue function. It has a much slower rate of elimination from the body; however, it is less easily absorbed from the digestive tract than radium."

Nothing was said of its effect on human lungs because experience with the synthetic element was so short-lived; in a year from then it *was* known.

FIFTH PHASE
A Man-Made Sun

• 1 •

This desert was age-old Indian Territory. The Pueblos had lived here nearly 2,000 years ago; before them Sandia Man, from caves in the Las Huertas Canyon, had cut arrow heads in antiquity to hunt the now extinct camel, horse, and mastodon—and had carved ivory from the horns of the American mammoth. Among the earliest hunters on the continent, they had lived up to 20,000 years before the big Army limousine came over the passes from Albuquerque carring the men seeking a site for the extreme sophistication of modern science, manipulation of nuclear physics.

Robert Oppenheimer, with Ed McMillan released from his radar work, rode with General Groves, on whom Oppenheimer had urged this mission. The work of bomb design was too fragmented and being set in different compartments, in various universities, threatened not only delay but security—and security touched a ready nerve in the general.

Oppenheimer argued further that a central cortex for building the bomb would have to be in some remote site, with isolated proving grounds. Experiments with normal high explosives would have to test atomic firing mechanisms and they had to be far from any community to avoid attracting attention. Yes, he knew the trouble he would have with academics, persuading them to leave their comfort and facilities to live and work in these arid wastelands to build the most terrible weapon in history. It had to be done; the scientist and the soldier were agreed on this. And so they came to the decision to find the site suitable for the secret atomic laboratory. It seemed preordained for Oppenheimer that they should inspect the old Los Alamos Ranch School for Boys near Otowi.

The rambling wooden buildings sat on the cone of a great

volcano that had died in pre-history and was now a wide mesa between two deep canyons. The snowy peaks of the Sangre de Cristo Range sparkled in the distance, the land sloped away down to the Rio Grande, and in the nearer slopes dozens of deep canyons radiated from rocky ridges. Oppenheimer loved this country; he owed it gratitude. When his life had been threatened with tuberculosis he had come to New Mexico and the biting, clear air had healed the suffering lung tissue.

This place, they agreed, met their requirements; true it had drawbacks in atrocious approach roads, lack of water and power, but Groves could swiftly handle those. The decision was instant, the action almost as quick. Within a month the army would own the Los Alamos Ranch School and the surrounding country, and it would enter the atomic empire designated as Site Y. Here would come the exotic products from Sites W and X, material that would be turned into the few pounds of metal to end the war. Site X was the huge Tennessee development under Black Oak Ridge—known first as the Clinton Engineering Works, later to be Oak Ridge. There a city of 50,000 workers was being created to service the great plants on 500,000 acres of land, in which gaseous diffusion and electromagnetic separation would take place. Here the work of Ernest Lawrence and his associate, Mark Oliphant, flowered with the so-called "race tracks"—doughnut-shaped magnets so powerful they could shift their own multi-ton bulks in their footings; magnets that had consumed almost all the national silver bullion from Fort Knox—valued at $440-million—for winding critical wire bands and forming special magnetic bars. In these "doughnuts" the uranium gas was beamed through the powerful magnetic fields that pulled the heavy atoms away from the lighter 235 isotopes to give an enriched product, a sticky mass on a metal collector the workers called "gunk."

Oak Ridge alone cost $500 million; and Site W would be as expensive and as complex. There, on a bend of the great Columbia River in Washington State, where the tiny hamlet of Hanford stood with its 100 souls, would rise the plutonium city. On 90 square miles of sagebrush and scant grass three huge uranium-burning piles would irradiate uranium metal slugs in neutron hailstorms to create Seaborg's element 94; by automation, operated by men shielded behind thick concrete, the slugs would be carried under water to the concrete canyons excavated in the wilderness several miles distant. There they would pass through the processes first devised by Seaborg and his team, to yield their cargoes of plutonium—freed from the lethal high radiation—for final purification into the fissile material for Los Alamos. A vast undertaking, Hanford needed services and power equal to that of a city of a million people.

As Robert Oppenheimer, now appointed director of the clandestine center on the high mesa, opened his task of attracting the greatest congregation in history of physicists and theoreticians to Los Alamos, Glenn Seaborg in the Chicago laboratories considered the ratios for production of plutonium metal that would make the atomic bomb possible.

Into his diary he wrote the steps that would lead to the flow from the remote-controlled caverns at Hanford to New Mexico:

"Reaction mass: One ton of uranium fissioned will yield 100 grams of element 94—plus 10 million curies of gamma radiation.

"Research Status: Removal of 94 from fission radioactivity *must* be by remote control.

"Step Three: Extreme purification of 94—especially from light elements which could induce unwanted neutron radiation—then, preparation of pure metal."

When the steps he envisioned were completed, the work of the three secret sites—W, X, and Y—held terrible implications for the people of two Japanese cities.

· 2 ·

The message came hidden in one of a pair of old keys on a small steel ring. Danish resistance men brought them ashore to Niels Bohr in Copenhagen from a night rendezvous in the Ore Sound. It was early March, 1943; the message was from James Chadwick. It was printed on a microdot, rolled into a fine hole drilled into the stem of one key. It brought challenge and heightened conjecture. It came at a time when Bohr, then recognized as the world's greatest theoretical physicist, the mantle handed down by Rutherford, felt all his strength and nerves strained by mounting threats, by Nazi violence, and terrorism.

Patrols roamed the streets of Copenhagen; armed guards stood at the doors of the Institute on Blegdamsvej; spurious students had been placed by the Gestapo to watch and report. The free speech and easy association that glowed in the memory of Otto Frisch had dissolved into a prison-like atmosphere. Other free spirits who had found a haven in the Institute in the prewar years had also fled, but there were still some vulnerable researchers, and only the immense stature and renown of Bohr stood between them and the concentration camps. His world standing had, so far, made him untouchable by the Gestapo, but there were ominous signs that terrorism in the occupation was increasing. Reports filtering through raised fears for his own safety as he barred German attacks on his staff. He had heard of the arrest of Paul Langevin in Paris; it could happen in Copenhagen. But, what if he left? What would happen to those who worked behind the shield of his reputation?

There was another matter burdening Bohr, and Chadwick's letter added to that issue: "I need not tell you how delighted I would be to see you again. There is no scientist in the world who would be more acceptable ... and I have in mind a particular problem in which your assistance would be of the greatest help. Darwin and Appleton are interested ... if you decide to come you will have a very warm welcome and an opportunity of service in the common cause."

It had been written on notepaper of the University of Liverpool; it was an invitation to escape, but more important it was *an opportunity of service in the common cause:* Defeat the enemy!

Bohr sealed the message in a metal tube and buried it in his garden at night; it stayed there until the war ended. *A particular problem:* that could only be uranium.

He had quietly discussed with his fellows and ruminated upon the question of the volatile isotope during the two and a half years of the occupation; he could not see how the 235 discovery could be used, for he knew no way of separation. And then his fears had been aroused when Werner Heisenberg, leader of the work on uranium at the Kaiser Wilhelm Institute, had arrived, unannounced, to talk about uranium metal made from the ore captured in Belgium and had also spoken of the heavy water from Norway. Unctuous, friendly, respectful—trading on his days as a prewar student of Bohr—he scouted this question:

Did Bohr think there could be a chance something other than a long-term project for power production might come out of uranium? Bohr did not know, and said he did not know, but Heisenberg's visit had left him very uneasy. Was this the secret weapon Nazi propaganda promised the world? Then, von Weizsäcker came and talked about military use of uranium; and now here was Chadwick's letter, speaking without the words. Were both sides dealing in the game for nuclear armaments?

Still, he rejected the idea. In his reply to Chadwick he said he had to help resist the threat to Danish institutions and protect the exiles; however, not even those duties, nor the danger of retaliation against his relatives, would keep him in Denmark if he really felt he could help: "I have, to the best of my judgment, convinced myself that . . . any immediate use of the latest marvellous discoveries of atomic physics is impracticable. However, there may come a moment when things look different. . . ."

Bohr was not alone in failing to see the path to the atomic bomb; the best of scientists in Hitler's Third Reich also missed the point. He endured a further six months of lawlessness, terror, and repression before the "moment" he had foreseen came to pass.

In September, the same Danish officer of the resistance who had carried the keys with Chadwick's message, came at night to Bohr's Carlsberg home, to warn him, "You are in great danger! You are to be arrested and imprisoned as an enemy of Germany."

His opposition to National Socialism was never flaunted, but he was well known as a Danish patriot. He asked the man, Captain V. Gyth, how certain this information was.

"We have our people inside the German staff. You must not waste time; it is imminent."

With his wife, his brother, and sister-in-law, all disguised as fisherfolk, Bohr was smuggled at night across the sound to Malmo in small fishing boats, experiencing alarms and escapes from German patrols on the way. Bohr's four sons and the rest of the family crossed to Sweden by the same method, but his infant granddaughter flew to Stockholm in a covered shopping basket carried by a Swedish lady diplomat. Escape ended Niels Bohr's internment and was the start of a bigger adventure.

In Stockholm, which his third son, Aage, noted was "swarming with Germans," a foolish official announced proudly that his country was host to one of the world's greatest brains—and so a cloak-and-dagger existence opened, with assassination a real risk. Then came a cable from Churchill's science adviser, Lord Cherwell; it said arrangements were being made for a flight to Britain.

A weekly flight of an unarmed, small bomber from Britain to Sweden was allowed by that country for diplomatic letters. On October 6, 1943, Bohr was stowed into the bomb bay of the two-seater Mosquito aircraft—a light, fast plane, built in part of plywood. Bohr was told that if the aircraft was attacked by German fighters, he would be dumped into the sea, with parachute and flares, and attempts would be made to rescue him by a fast patrol craft.

Bohr's flying helmet was far too small for his massive head, the earphones did not reach his ears, and he didn't hear the instructions from the pilot to breathe oxygen. He fainted, and did not recover until the plane lost height to land in Scotland. His son, Aage, who was to be his secretary and assistant, made the same journey a few nights later. He had no such trouble.

Father and son were lodged in a Westminster hotel and allocated rooms in the offices of the Tube Alloys Directorate in Old Queen Street. There, Bohr heard from Chadwick of the Peierls-Frisch concept of a critical mass of uranium isotope, of gaseous diffusion for 235 separation, of the Fermi pile for plutonium production. In return, he had grim news, leaked from Germany. Big efforts were being made to restore the water plant in Norway, and an industrial plan had been stepped up to provide uranium metal for an experimental reactor in the grounds of the Kaiser Whilhelm Institute, to be built by a group under the guidance of Heisenberg and von Weizsäcker.

Anxieties were revived in Britain and America. Bohr wondered whether the German physicists had found element 94 and the new path to the atomic bomb. In his notebook, Aage recorded his father's growing fears for the free world.

They went to join the Manhattan Project, to be astounded by its immense scope and sophistication, and to be confused by the

American security system, which changed their names to Nicholas Baker and his son James. Among the irreverent at Los Alamos, it was inevitable the great Dane would be known as "Old Nick."

Otto Frisch spent the summer months of 1943 working in the comparative peace of Oxford. The pride of British uranium work was then the gaseous diffusion plant to win pure supplies of isotope 235, and he had ideas for building a new instument for routine testing of hexafluoride, the gasified form of the element. It was a sojourn from his post in Liverpool; there, the savage bombing had driven him from his boarding house, and days at the laboratory were often spent patching windows with cardboard. When summer was almost done, he returned to Liverpool, dreading the thought of the coming winter and the continuing blitz.

One day, Chadwick halted him in the corridor and asked, "How would you like to go to America and work there for a while?"

Surprised, Frisch could only say, "I think I would like that very much."

"It would mean, of course, you would have to swear allegiance to the King, forsake other national bonds—in fact, you'd have to become a British citizen."

Frisch beamed; four years had passed since he was given a berth in Birmingham by Oliphant, four years as an exile and an enemy alien, barred from national projects, but allowed to work on atomic energy. He told Chadwick, "I would like that even more."

He was to be one of a British team to join the Manhattan Project. He was hurried to London, where he swore an oath, obtained naturalization papers, won exemption from military service ("otherwise you'll be a deserter," they told him), was handed a passport, and sent to the American Embassy for a visa—all in one day. Then it was back to Liverpool to board the liner *Andes*, being used as a troopship.

The British team was hand-picked for outstanding expertise and to supplement the vast American effort. Along with Frisch it included Peierls (who was to succeed Chadwick as leader), Oliphant, who was to work with Lawrence, Egon Bretscher, Titterton (Oliphant's former student), Placzek, William Penney, Joseph Rotblat; among the rest was a German exile named Klaus Fuchs.

Edgar Sengier also crossed the Atlantic, but in the opposite direction to the British mission's, and under far less pleasant conditions. He was put aboard a troopship crammed with American soldiers bound for the Normandy invasion.

His mission was to ensure that America had near-total control of the known uranium reserves, and his journey had started with two plainclothes agents of the Manhattan Project escorting him to the docks in New York. As Sengier later recalled, "It was six in the evening. It was all so secret they prevented me saying goodbye to my wife . . . she had the impression I was being taken to Sing Sing."

The crossing ended with similar secrecy. At Liverpool, he aroused much suspicion. There was discussion as to whether he should be returned to America aboard the ship, or be quarantined, when an imperious British colonel intervened, waved the dock guards aside, planted the disgruntled mining tycoon and his bags in a jeep, and drove through the rain to London.

In New York, Sengier had been under pressure from Groves for many months to sign away future extraction of uranium to America. In mid-1943 Groves had been given authority by the President and the U.S. Military Policy Committee to "allow nothing to stand in the way of achieving as complete control as possible" of the world's major uranium resources; and, with this end in mind, he argued with Sengier that American experts should reopen the African mine.

Resolute in his view that he should keep control on the Shinkolobwe deposit, unwilling to concede loss of ownership, Sengier argued that reopening the mine was prevented by duty to his company, one-third of the stock of which was still owned by former British interests in the old Tanganyika Concessions Ltd.

Sengier also had a keen eye for the future of uranium and suspected the real reason for Groves' feverish drive for ownership of the major uranium supply. In mid-1943 Groves had a secret survey made of known world resources of uranium; in this, the Katangan ores stood out as critical to all the military planners had in mind for the future, so superior in richness and quantity that, even a year later when material flowed from Canada and Colorado, Shinkolobwe still provided four fifths of all fissile fuel in America.

Once Groves accepted that Sengier's attitude would not change, he hatched a plan to involve the British into putting pressure on the Belgian government, then in exile in London. So it was that, at the invitation of the exiled government, Sengier was on his way to a conference to negotiate the release of the capacity of the uranium mine in Katanga to feed the world's first atomic arsenal.

By that time the original shipment of ore had been run down by the many hundreds of tons sent from the depot at Seneca to the Port Hope refinery for conversion to high-grade oxide. Much of this had already been converted to the metal slugs to feed the uranium-burning reactors for plutonium production. Some had gone to be mixed with fluorine for hexafluorine gas to feed the giant magnet systems called "race tracks."

In the view of many scientists there was then more than enough uranium in America. Some of the leading minds had argued against the scale on which Groves was building; given the material and backing, they said, a compact group of the right men could build a bomb, or two, in a few months. Groves had a broader view, not openly stated. He wanted the complex work at Los Alamos to "proceed at a comfortable pace." The scientists saw their great effort aimed at ending the war with one, or two bombs. Groves saw much further; military planning was geared for 25 bombs in the first year, enough strike power to devastate a major city every two weeks. And Groves, Vannevar Bush, and the Military Policy Committee looked well beyond that.

Los Alamos, with the greatest concentration of scientific talent in the world, aimed at nuclear domination beyond the conflict and a new era of power production from clean, silent palaces of electrical generation. America was to be in charge when peace came. The role Sir John Anderson had seen for the two nations, as global policemen, America would fulfil, alone.

The government-to-government discussions in London between the American ambassador, John Winant, Sir John Anderson for Britain, and the Belgian exiles, resulted in agreement— but not what Groves wanted. His desire for a 99-year contract for

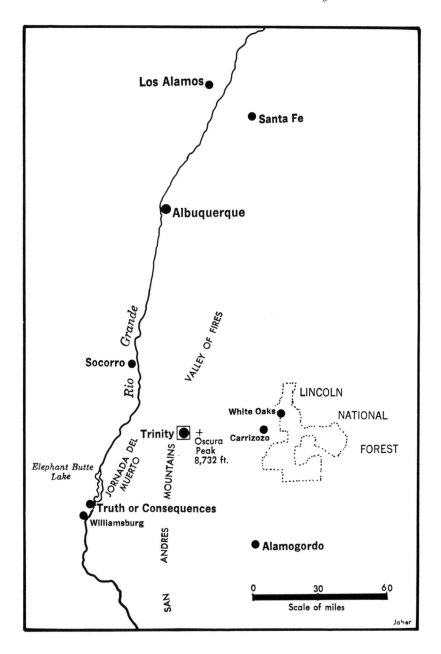

Area of New Mexico showing Los Alamos and also Trinity, the site of the first atomic explosion

first claim and use on the Shinkolobwe deposits was rejected; Sengier was still unwilling, but finally agreed, under pressure, to opening the flooded mine and supplying a further 1,720 tons of ore to a new Combined Development Trust Fund, in which America, Belgium, and Britain were partners, and a right of first refusal for a period of ten years afterward of all excavated ores at cost, plus a "fair profit." Groves was chairman of the new board of trustees; and, when the Allied armies liberated Europe that summer, he was able to bring more of the Congo ores, captured from the Germans, to the United States.

• 3 •

The effectiveness of the immense atomic empire built by mid-1944 was still totally dependent on solutions to remaining mysteries in the uranium atom and its offspring. How they would behave in full-scale fission was vital to the success of the whole vast Manhattan undertaking. Scientists at Los Alamos still had to find the path to management and manipulation of the explosive atoms.

The "gunk" that came from Lawrence's magnetic "race tracks" and the plutonium from Hanford had to be explored for their chemical and physical snags; and the safest, quickest, most effective ways of turning that material into dense metal had still to be found. The real critical mass of 235, or plutonium metal, was no more than scribbled equations. Nobody could actually measure, record, and be certain of the time lapse between the first neutron strike and total fission, or of the pressure generated from the splitting nuclei and the outward thrust, nor say how much of the metal was actually fissioned before the heat vaporized the remainder and the explosion blew it into surrounding space.

Yet, the world's pre-eminent scientific brains were assembled here by Oppenheimer. In fact, they were employed by the University of California because they refused, as heads of divisions, to be majors in uniform working under direction of "Lieut-Colonel" Robert Oppenheimer.

In the strange world of Los Alamos, the word "fission" became a part of the casual jargon for the nature of the most complex atom and was reduced to "fish"; numerals were inverted to gain simplicity and to guard secrecy—common uranium 238 became 28, isotope 235 became 25, and Seaborg's brainchild, element 94, became 49. These were building blocks for the objective spelled out for all newcomers who traveled the road from Santa Fe, crossing, as William L. Laurence wrote, "... from yesterday to tomorrow." Through the mountain passes into Los Alamos the convoys of men and machines poured; cyclotrons from various universities, with spectrographs, and all the impedimenta of the nuclear-metallurgical world were brought to the mesa. In his introductory lecture Robert Serber, told newcomers, "The object of the project is to produce a practical military weapon in the form of a bomb in which the energy is released by a fast neutron chain reaction in one or more of the materials known to show nuclear fission ...."

The direct energy release per atom in fission was 170 million electron volts—more than 10 million times ordinary chemical combustion—and from this they could teach, "... one kilogram of 25 'fished' would equal the power of 20,000 tons of TNT."

At Los Alamos, they could now count those atoms like marbles in a bottle: "Since in one kilogram of 25 there are some 50-million-million-million-million nuclei of uranium atoms, it will require about 80 generations to 'fish' the whole kilogram."

From the intake of the first neutron in the first 235 atom and the release in the explosion of the atomic core of two or more neutrons, it would take 80 such steps to generate the explosive force of 20,000 tons of TNT. It was a simple chart to a bomb—except for the unknown vagaries. At what lightning speed would the fission of 80 generations of nuclei take place? There *could* be a point in that flash of time when the released energy would be blasting the material outward, allowing neutrons to escape without seeding fresh atoms.

They could say, "The speed of one-million-electron-volt neutron is near to one-thousandth-millionth of a second over the distance of a centimeter, and the mean time between fissions is about a hundred-millionth of a second, given a free path of up to 13 centimeters."

Thus, it was argued, it would be only in the last few fission generations that enough energy would be unleashed to spread the rest of the active material outward. With the release of infertile neutrons, the energy system could be disrupted early; so, like sparks from the plugs in an automobile engine, the neutrons *had* to cause fission to be effective. This meant that a sphere of explosive metal of 235—or plutonium—had to be much larger than a single kilogram to maintain high neutron density, to prevent the sub-atomic particles escaping into the outside world. And here, there came into play a Peierls-Frisch notion, first suggested in the memorandum which led to the foundation of the British M.A.U.D. committee. This was the idea of the "tamper," a surrounding shield of other metal, possibly beryllium or common uranium, which would *reflect escaping neutrons back into the heart of the bomb!*

They reasoned that, with a tamper of normal uranium, the diameter of a critical mass of 235 metal might be around 15 centimeters, the weight about 15 kilograms. Five kilograms of plutonium would be needed since this was judged to have a faster fission ability than the uranium isotope.

In the final outcome, the estimates first written by Frisch and Peierls for a critical mass of 235 were to prove fairly accurate. American nuclear weaponeers would talk, then, of "the five-kilo bomb."

While chalked equations could go no further than estimate the critical mass in early 1944, theory could, nonetheless, forecast the horror this appalling "practical weapon" would unleash. More than a year before the nuclear shock front was first measured, they were able to predict the mechanics of the blast waves and their ferocity: "In a bomb containing five kilograms of active metal one could expect a destructive radius of about two miles from the kind of device we hope to make."

Devastation would occur over an area of more than six square miles—from a few pounds of metal. The calculations went on into an aspect of combat: "A very large number of neutrons is released in the explosion. One can estimate a radius of around 1,000 yards about the site of the explosion as the size of a region in which the neutron concentration is great enough to produce severe pathological effects.

"Enough radioactive material is produced so that the total activity will be above one million curies, even after 10 days. Just what effect this will have in rendering the locality uninhabitable depends on how the matter is disbursed by the explosion . . ."

Uranium fission by-products were also to be a weapon of war. Not just a single burst of concentrated metal exploding above a city but as well, copious amounts of lethal radioactive waste scattered wholesale through the countryside; invisible walls of death to confront Allied soldiers when they landed in France to start the liberation of Europe. General Groves saw it as an act of which the ruthless Nazi leaders were capable. Rather than surrender a yard of territory, Hitler could order Field Marshal Rommel to use such indiscriminate methods as easily as President Roosevelt would unleash the atomic bomb over German cities.

There was further hidden cause for his concern. Paradoxically, the fear came from the growing certainty the enemy had *not* started to build atomic bombs, a prospect that had seemed very real in 1943, based on rumors of work with uranium metal from the captured Belgian ores being used in experimental piles in Berlin and elsewhere; so strong had been the rumors, senior science advisers had urged him to alert the American public how they might come under nuclear attack from Germany. He had landed agents in Italy and had men working, through British intelligence, in London; in America, the European exiles had been questioned—Frisch, Wigner, Teller, Fermi, and Bohr when he arrived—on the identities of Third Reich scientists most likely to work on building a bomb. From underground information, from German newspapers, and through neutral countries, they had built a picture, matched with aerial photography; and nowhere was there a sign of the kind of massive development Groves knew to be essential to production of fissile metal.

But Conant, Urey, and Compton did a study of the probabilities and estimated Germany could be making plutonium in reactors in various places, that the bomb prospect might be very small, but radioactive missiles were possible.

Very few people knew of the danger. Groves wrote, "It was one of the biggest risks taken by the United States during the war. It gave the few who knew about it some bad hours."

The possibility could not be avoided; it could only be countered by instant detection—and out of that came Operation Peppermint.

Groves consulted General Marshall; a letter from the Chief of Staff went to General Dwight Eisenhower at his SHAEF location in Britain: ". . . certain materials might be used against your Armies in a landing operation. The matter is of the highest order of secrecy."

Groves had specialist officers trained to operate Geiger counters on seek-and-detect assignments in Europe; men of the U.S. Chemical Warfare Service were equipped with sensitive detection instruments, the Signal Corps had electronic experts standing by to service them; the whole Medical Corps in Europe was alerted for telltale signs of such an invisible assault on the fighting troops and were told the symptoms to watch for should men fall suddenly ill for no apparent causes. Also, X-ray units and official photographic sections were alerted to watch for film which became inexplicably fogged.

When General Eisenhower reported to Washington that plans had been laid for Operation Peppermint, Groves felt much easier. "Nothing more could be done, except to pray that we had not made a mistake."

In June, 1944, his anxiety about Allied fighting men being embroiled in radioactive war vanished with the successful Normandy invasion—and was replaced by a new and daunting prospect. There was serious doubt plutonium could be used in a bomb.

In the first batches of the transuranic element which came from Devil's Canyon there was troublesome extra activity, unsuspected alpha radiation that would threaten premature fission in a plutonium weapon; this meant that the huge Hanford plant, built with immense effort and speed at a cost of $400 million, might be useless for the war effort. The President's hopes to strike at Germany with the first nuclear explosions would depend entirely on the agonizingly slow dribble of enriched uranium, which the 'gunk' workers were scraping in tiny amounts from the electrodes of Lawrence's giant magnets at Oak Ridge, and a stepping-up of the uranium gaseous diffusion project.

The Manhattan commander had gloomy visions of facing Congress when the war was finally done and trying to explain how $2,000 million was spent on a weapon that was never used.

· 4 ·

The mesa was white with Christmas snow when Niels Bohr and his son arrived at the Los Alamos Laboratory. Bohr came into a situation that was afterward described as "a seething crew of genius." Direction and administration by army officers; constant arguments with builders and technicians on facilities, many of the scientists aware that their university pay was around half of that being earned by the artisans; the seclusion of the site and lack of adequate services; and, most of all, lack of understanding of the scientists' approach—all of these factors led to "relentless strife" between factions.

When he was appointed, in May, 1943, to head the work at Los Alamos, Robert Oppenheimer had been under pressure from General Groves to accept an Army commission and put the work on a military basis. He had consulted Isador Rabi and Robert Bacher, then working with Bainbridge at M.I.T., and had asked advice on the management of a big defense laboratory. He had been advised, in strongest terms, "Research must be kept free from Army control. Let the Army do the bookkeeping, run security and supply, but always keep the research under civilian control."

That advice laid one base for a head-on struggle with General Groves; there was also pressure from men heading Groves' security corps to make Oppenheimer's role less powerful. The earlier associations with Communists, the spiritual impact of the Spanish Civil War, which gave him the status of a "fellow-traveler," were a constant source of concern to these officers; and though Groves gave him clearance he was never free from surveillance. When the war was over it was revealed publicly how he had been harassed, his mail opened, his telephone tapped, his movements closely watched, how he suffered constant questioning—all this amid the harrowing task of placating his growing ranks of scientific elite, smoothing the "seething crew of genius," shielding them from the impact of the military mind.

Then Oppenheimer gained a powerful crutch. With delays and uncertainties adding to the troubles from other directions, the intellect and vision of Niels Bohr was a panacea. The official Los Alamos history reads, "His coming had a very healthy influence on research. He came at the right moment ... the innumerable small problems which confronted the physicists had led them from fundamental problems of the bomb. Study of the fission process had been neglected, obstructing reliable predictions of important phenomena ... Bohr's interest gave rise to new activities ... which cleared up many questions left unanswered before."

His knowledge was felt strongly in the study of the metals to be used as a tamper to the bomb, and these included platinum, gold, iron, common uranium, and beryllium—yet another of the elements Klaproth had discovered.

The Los Alamos report adds, "Last, but not least, his influence on morale ... must be mentioned. It went further than having the great founder of atomic research in the Laboratory, beyond the stimulus of his fresh suggestions. He saw administrative troubles in a better and longer view than many enmeshed in them. His influence brought about more consistent co-operation with the army in pursuit of the common goal. And what can be least overlooked—he gave everybody ... some of his understanding of the ultimate significance of the control of atomic energy."

Bohr turned to seeking alternate methods of bomb assembly, to the velocity of neutrons, and to a unique method of firing the plutonium bomb—an attainment that saved the Manhattan Project from failure.

The problem came with the discovery of a new plutonium isotope.

From the start of the atomic bomb concept, solution of the problem of how to achieve a critical mass of uranium-235 in the microsecond to avoid premature fission, had rested on a gun-barrel assembly; a smaller projectile of uranium metal would be fired by high explosive charges along the barrel to collide with a subcritical assembly of uranium. The total volume would then be above critical mass and the chain reaction would occur instantly. The original speed of the uranium projectile of 6,000 feet per second had been brought down, by calculation, to between 2,000 and 3,000 feet; but there was still a problem, simple to state—a light-weight gun had to fire the uranium slug fast enough to cut out the risk of a stray neutron causing a premature fizzle. There was more at stake than the chance of the bomb being a flop. Dropped over enemy territory and failing to fire properly would do two things: it would give away the bomb secrets; and the enemy would have the gift of critical, priceless, fissile metal.

Other solutions were suggested, and among these was the system to be called "implosion." It was an act against nature. So long as the gun-bomb assembly was feasible, implosion had not had much currency—until the first solid samples of heavily irradiated plutonium arrived at Los Alamos from Oak Ridge.

Radiation activity in plutonium was obviously greater than in uranium-235. In the previous summer (1943) crucial questions arose on the numbers of neutrons plutonium emitted under fission, and these had to be answered before the flow of material came from the big Hanford reactors. To meet this situation, Seaborg divested his Chicago laboratory of a sample of 200 millionths of a gram which Art Wahl and others had extracted from uranium irradiated with the Lawrence 60-inch cyclotron in Berkeley. An armed escort brought this to New Mexico.

The plutonium was put through a series of tests by Robert Wilson, on a cyclotron installed at Los Alamos, and revealing data were obtained of explosive potential. When the analysis was done, Wilson heard that Seaborg, with his wife Helen—the secretary to Ernest Lawrence—was on vacation in Santa Fe and would soon be going home.

Always looking to save time, Wilson cut the cumbersome Army regulations and early one morning with the protection of a high-powered rifle, took the plutonium in his car to Santa Fe. It was 6 A.M. in the all-night Blue Ribbon Restaurant, and Seaborg and his new wife were early at breakfast, before taking the Santa Fe express back to Chicago. Wilson stalked in, eyes roaming the big room, rifle in hand; surreptitiously, he handed the small pack to Seaborg, who then carried the world's largest sample of plutonium in his briefcase aboard the train, all the way back to the Chicago campus. These first tests on element 94 at Los Alamos aroused no crisis; this came when the first samples of reactor-produced plutonium arrived in late spring, 1944.

The new samples had been extracted from uranium more heavily irradiated than any in history. In the reactor chain reactions, neutrons had continued to smash into uranium-238 atoms which had already been neutron-seeded, into the cores of atoms which had decayed through neptunium into plutonium. It was discovered at Los Alamos these atoms of plutonium had an additional neutron; there was an isotope, plutonium-240. This endangered the bomb project. The plutonium isotope was a strong emitter of alpha particles, and that meant a fresh neutron background. It was a threat that demanded gun velocities far beyond any yet achieved.

The risk of stray neutrons pre-detonating the plutonium bomb was now enormous; it would endanger the bomb assemblers and the air crew carrying the weapon to its destination. The plutonium gun-bomb was abandoned. In its place came the double-bomb, the implosion system that Niels Bohr, together with its original advocate, Dr. Seth Neddermeyer, proposed and supported. It was the only chance of making Hanford and its plutonium useful in the atomic bomb program.

Oppenheimer called for a full-scale attack on the implosion studies: "Throw everything in; throw the book of knowledge at it!"

Applied science, technology, metallurgy and chemistry, and a hundred skills were bent to defeat the threat of plutonium-240. Month after hectic month the work went on, even as Allied troops fought across France, liberated Paris, subdued the flying-bomb bases in Holland, and faced the last savage fling by von Rundstedt through the Ardennes—at Christmas, 1944, the sixth Christmas since fission.

Kenneth Bainbridge knew at once he had come on the site for Trinity when, in August, 1944, he drove from the White Sands range and through Mockingbird Gap, north of Little Burro Peak. This flat area of desert in New Mexico met Oppenheimer's requirements—for observation, for minimizing risks from wide scattering of poisonous plutonium from a premature firing. The order was that not one Indian was to be displaced, and there were no Indians—nobody—and the site could be seen from more than 20 miles away. No other place they had considered met the requirements like this tract; not the Channel Islands of California, not Padre Island, or the Texas sand barrier, not the Mojave Desert.

This was part of the bombing range, with Alamagordo air base 60 miles south. There was a large, wooden bombing target set in circles of ploughed land, and as he looked he heard the throb of engines and saw a flight of B-17 bombers. He was instantly apprehensive. The air force had been warned there was to be an inspection of this area, and here were the rookie pilots, the open bomb bays black rectangles on the bellies of the planes—and now the bogus bombs, of sand and a few pounds of high explosives, were crumping into the area of the wooden frame. From that day on, until the dawn when the first atomic explosion was fired, Bainbridge feared the coming of the uninformed trainee pilots. This area of arid land, 18 miles east-to-west and 24 miles north-to-south, would be off limits, by order; indeed, as it had been declared for this first day! Marking it on his map, he saw it sat in the middle of the long valley which the Indians had named Jornada del Muerto—the Journey of Death.

Bainbridge was appointed to take charge of the Trinity test. Almost a year of preparation awaited him, once he laid out Zero point and planned the location of the bunkers, north, south, and west, at 10,000 yards; he persuaded Groves to build temporary roads and had 500 miles of cables laid and then turned the talents available to controls, remote firing, monitoring, electronic devices that would measure the energy release. He had no compunction in helping to launch the apocalyptic atom on the enemy; he had "a somewhat bloodthirsty viewpoint on the war." He had seen the flow of persecuted humans coming out of Germany, and he had endured aerial bombing in Britain; he had firsthand accounts of Guadalcanal, and Crete, and the extermination of the parents of fellow physicists. Faced with Hitler's Thousand Year Reich and the war getting "dirtier," he was "glad to help get there first with atomic weapons."

Meanwhile the Kaiser Wilhelm Institute was already bombed into ruin. The bombing of Berlin also drove the core of German nuclear physicists and chemists to repeat the action of the French team in 1939; they fled to the south of the country.

Heisenberg, Weizsäcker, Harteck, and others dismantled their crude uranium reactor assembly at Dahlem and, with several tons of uranium metal, went down to Württemberg, to the redoubt area of final Nazi resistance. Hahn, who had filled the war years following the devious chemical paths of fission by-products, went with them; and so did the Germans' major stock of uranium. Manhattan Project agents, ferreting out fragments of information as they advanced across Europe with Allied forces, found there was no longer any question of a race with Germany for the atomic bomb. They had talked with Joliot-Curie in Paris and had gone on to interrogate Union Minière executives in Brussels; however, there was still cause for concern, there was still a risk of radioactive retaliation in the last desperate hours of Hitler's Third Reich. There was a huge amount of Shinkolobwe uranium unaccounted for somewhere in Germany.

In Brussels, Groves' agents learned some of the details. There, Gaston André had held the fort for the absent Sengier during the occupation years; nobody has learned from him—not to this day—the true size of rich uranium concentrates from radium extraction in the waste heaps at Oolen when the Germans invaded Belgium in May, 1940. André, however, did report despatch of increasing amounts to German chemical firms during that time; also, that 1,200 tons of fissile treasure had been looted by General Schumann's orders and shipped to a secret destination.

Groves' agents finally discovered where the uranium had been secreted. The nearest town was Hechingen, between the French drive to the east and the leading Russian troops driving west, just north of the Austrian border.

A special task force was formed; the operation was called Harborage. It went into redoubt territory, came under sporadic sniper fire, and it found the haul of Katangan uranium in a fragile shed, in thousands of deteriorating wooden barrels. The troops and officers, and some 200 truck drivers, toiled to shift the uranium bonanza; but first, new barrels had to be provided, and a local German firm was ordered to fabricate 20,000 barrels in two weeks.

The pace of this frantic action, its objective hidden from the French and unknown to the advancing Russians, puzzled the men who re-packed and loaded the barrels of yellow ores. One Allied officer attached to this venture was Sir Charles Hambro, a member of the British banking family, and this led to a rumor that the yellow ores contained rich deposits of gold; others thought it had to do with some secret ingredients for brewing beer.

The great consignment of uranium was trucked across Europe; some went to England by air and some by sea. Groves was delighted with the news and wrote this memo to his chief of staff, General George Marshall:

"In 1940, the German Army in Belgium confiscated and removed to Germany about 1,200 tons of uranium ore. So long as this material remained hidden, under the control of the enemy, we could not be sure but that he might be preparing to use atomic weapons.

"Yesterday, I was notified by cable that personnel of my office had located this material . . . and that it was now being removed to a place of safety. . . .

"The capture of this material, which was the bulk of uranium supplies in Europe, would seem to remove definitely any possibility of the Germans making any use of an atomic bomb in this war."

All of the Congo ores captured in Europe and retrieved from Germany were shipped to the United States to feed the nuclear weapon development that followed the defeat of Germany's last ally, Japan.

· 5 ·

Frisch was quartered in the old wooden building of the ranch school, which was called the "Big House," and, with other bachelors, ate his meals in Fuller's Lodge. He liked this life, and he wrote to his parents, then in Sweden (although he could not say where he was). "I am at an ideal holiday spot with excellent facilities for work. . . ."

He said he had never been with a more interesting concentration of people; he felt he could wander at evening into any home, at random, and find congenial folk enjoying music or engaging in stimulating conversation and debate. It was a tranquil picture for the grim purpose of the complex and the long, strenuous days of preparation.

Specifications of the atomic weapons couldn't wait for precise definition of the critical mass of metals. The chemistry and metallurgy sections under Joe Kennedy, had still to devise new methods for turning the plutonium from Hanford and the uranium-235 from Oak Ridge into exotic metal of the highest purity, and then prepare the volatile material in shapes which the scientists could safely use.

Until then, Oppenheimer had to drive his teams toward an understanding of specifications that would be close enough to eventual requirement as to be capable of adjustment. He put Frisch in charge of a special critical assembly group and housed them in a spacious laboratory built in the deep Los Alamos Canyon and named "Omega." The size of cores for the atomic bombs would be determined here.

The work done in the canyon entailed experiments, known as "Dragon" and "Lady Godiva," which were to bring Los Alamos within two seconds of disaster and Otto Frisch closer to death than he had ever been.

With the increasing flow of fissile material now predictable, urgency lent weight to all they did, and the decision on bomb design had to be final. The implosion method of firing would be used for the plutonium bomb, while the Frisch-Peierls gun-barrel system still stood for the uranium bomb. In the canyons, tests were made of muzzle speeds with three-inch artillery, Hispano guns, and the lightweight barrel planned for a single firing. In the case of plutonium, the work took on what the official record called "an aura of grim desperation."

Bainbridge was drawn from work on Trinity preparations to work on the solution to the difficulties in constructing the plutonium bomb. This was to be a shell of high-energy chemical explosives fired simultaneously by electronic detonators, all set inside a spherical steel container. The purpose was to blast *inward* the metal segments of subcritical size to form a mass (or volume) of plutonium that would explode. And this had to be done within the millionth of a second needed to avoid premature fission—each segment to be fired inward at *precisely* the same time.

In the middle of this complex work, however, the question was asked, "Since this will be a bomb dropped by parachute, not falling free like normal bombs, because we want it to explode for maximum effect a thousand feet or so above the city, what, then, will happen if Japanese fighters get close enough to pump cannon fire into the falling 'Fat Boy'? "

With Bainbridge's experience of the Trinity bombing range, it presented a hair-raising vision of some "budding young bombardier," oblivious of the secret bomb project, mistaking the steel tower at Trinity, supporting the "Fat Boy" bomb (as the plutonium bomb was called), for his target area. He made a mental note to press for radar-directed anti-aircraft guns to fire warning shots in the path of such planes. It also meant the "Fat Boy" casing would have to be thoroughly tested.

The metal-casting team made a mockup of the implosion bomb, complete with high-explosive shell, but with a dummy core. Bainbridge recalled, "The idea of the test was to shoot 20-millimeter shells into the spherical casing to see what Japanese fighters might do to a real plutonium bomb. . . ."

The mockup bomb was placed on the mesa and the Army gunners lined up their cannon from several hundred yards. The stream of shells smashed into the sphere as Bainbridge watched from a bunker; after he halted the firing, the round steel casing lay inert, and, with a companion, he left the bunker to inspect the damage but quickly retreated when they saw smoke curling from some of the shell holes in the casing. They let it sit for a few hours, still smoking, before Bainbridge again went to make an inspection. He found the shells had pierced the steel and the high-explosive coating; by good luck, not one had struck a detonator. The Army demolition men then put charges under the "Fat Boy" dummy and blew it out of existence.

The day the uranium-235 metal was delivered to Los Alamos Canyon was a landmark in the career of Otto Frisch. It was early April, 1945, when an Army sergeant dumped the small boxes of packed metal on the bench at the Omega critical assembly laboratory. Frisch opened the packing and handled the first 235-metal made. Here, at last, was the metal he had discussed with Peierls during the blitz in Britain, the substance Bohr saw as the key to fission, the metal of the isotope he had tried vainly to isolate with his simple Clusius tube in 1940.

The new metal was beautiful to him; it gleamed silvery and brilliant, and he held a block—there were hundreds of little oblong blocks, measuring one inch by a half-inch—and admired its gloss and glint. It was quite heavy; a rapid mental calculation: four blocks to a cubic inch, a cubic inch around 70 grams, and eight kilograms would make a sphere of roughly six inches in diameter—a ball of metal he and Peierls had predicted would make an "enormous explosion."

Gradually, the silver grew dull and changed to a blue sheen that deepened to a rich plum, nearly purple. The oxygen in the air was reacting with the surface atoms of the metal; it was a small reminder of the intense volatility of uranium-235.

Some innate sense of ownership, from his long association, possessed him. "I had the urge to take one; as a paperweight, I told myself. A piece of the first uranium-235 metal ever made. It would have been a wonderful memento, a talking point in times to come."

The urge was quickly rejected. Every minute speck of this material was priceless—the end result of America's $2,000-million investment in the uranium project. His task was immediate—to use these small blocks to build a sphere almost to the point of critical mass, to establish by practical test how much of this metal was needed to create an efficient atomic explosion. On what he did with these little metal blocks depended the final assembly of the world's first uranium weapon; how much would go into the projectile fired down the gun barrel and how much metal would have to be in the target so that, together, they formed a critical mass. Accuracy in the first instance was super-essential; this metal was in very, very short supply. There would be no test firing, as with plutonium because the "gunk" from which it was smelted was not yet flowing fast enough from Oak Ridge.

His was a dangerous task, not to be hurried; yet, pressure was coming from Washington—after all those lost months!

Frisch had proposed to the Director, Oppenheimer, and the head of the theoretical work, Hans Bethe, "Look, I've had a brain wave—let's see how near we can go to the limit which your theoretical people have set for criticality. It will test your doubts and establish accuracy for you. We can go cautiously and maintain control, and when it's done—you'll know! You will then be confident in what you have to do."

He didn't think they would accept the notion; he was surprised when, on April 5th, Oppenheimer told him to go ahead. From that day on he set up the first experiments with near-critical masses that became known as the "Dragon" series. One of the leading Los Alamos personalities was moved to comment that Frisch was "tickling the tail of a dragon."

At first, they used a sample of uranium-235—hydride, the pure uranium extract mixed with hydrogen, not yet made into metal but quite capable of critical fission; in fact, at one time it had been suggested a 235-hydride bomb might be feasible.

Frisch and his team built a wooden frame ten feet high. Within this structure ran a four-cornered chute of aluminum; a uranium slug—two inches by six—was dropped down this. Halfway down the shute the slug fell through a ring formed from small blocks of hydride. When these two pieces were together for the microsecond the slug was passing through, there would be a burst of neutron activity, and this activity would increase the closer the two pieces, combined, came to the volume of the critical mass. The method of measuring this volume was more intricate. Frisch arranged a refinement of devices he had used with Titterton at Birmingham. He had the usual Geiger counter with its mechanical clicking, and there was a row of five little neon lights, each scaled to flicker with mounting intensity of bursts of radiation; from one to five, they indicated increasing levels.

A hundred or so times a day the uranium slug would drop through the hole; each time the amount of hydride in the middle of the frame would be increased slightly by adding small blocks. Sometimes they would experiment with tamper material, such as beryllium oxide, set around the uranium to measure the increased effect of reflecting escaping neutrons back into the uranium mass.

Then the first metal was delivered to the laboratory. The work of establishing the accurate critical mass changed pace and became more demanding.

To determine the exact fission capacity of the metal, Frisch decided to build a small sphere, as near as he could with oblong blocks, and to take this, very carefully, from subcritical size to the point where instrumentation would indicate fission chain reaction was imminent.

There was to be no tamper. A small neutron-producing source was set inside, and the reaction to be measured was that of naked uranium-235 metal. Frisch called it the "Lady Godiva" experiment.

There were more than 400 of the small blocks to be laid around the neutron source; it was a long, tiring task, placing one block at a time. "Like a child at play," he proceeded steadily with the mechanical chore.

He finished the eleventh layer. The Geiger counter was clicking fast—the five little neon lamps were winking at different speeds, registering the flow of activity. Piece by piece, he placed the twelfth layer, bending over, his white coat loose around him, cautiously adding one small block after the other.

Suddenly, one of his assistants stepped forward, pulled the instrument away from the metal assembly, saying, "The counter has stopped working."

For Frisch this was a moment of distraction while placing another block on the top layer; his rules were firm against changing the circumstances of an experiment. He shouted, "Don't move that. Put it back, for God's sake . . ."

In that second, the terrifying truth flashed into his mind. He could sense, more than see, catastrophe being signaled—the counter could not register the surging flow of radioactivity. From the corner of his eye he saw the bright neon lights staring at him, no longer flickering, their gases swamped by the swiftly mounting flood of neutrons—from this mass of metal under his *bending body!*

Frisch knew his white coat was a cover for "Lady Godiva"; the material and his torso were reflecting escaping neutrons back into the assembly. It was fast rising to an explosive situation. With a swing of his uplifted arm he swept most of the top layer of blocks on to the floor—and straightened his back.

He had come closer to assembling a critical mass of uranium than any man before. The radiation was so fierce the neon lights still stared at him in warning. But, what unparalleled opportunity! Here was the chance to calculate the exact volume of the bomb core!

He made readings of the radioactivity and determined that he had suffered a neutron saturation equal to a day's total permitted dose, within the space of two seconds. He knew, then, he would live.

The rate of radiation of neutrons at the instant he had swept the blocks away had been multiplying by a factor of one hundred, *every second of time!* The arithmetic showed near-catastrophe for both Frisch and Los Alamos, but it gave the data for which the engineers waited.

That night, April 12, 1945, at dinner in Fuller's Lodge, Frisch said little of his escape from death; another event of that day took precedence. At Warm Springs, Georgia, President Franklin Delano Roosevelt had died from a cerebral hemorrhage at the age of 63. He was succeeded by Harry Truman.

Next morning, Frisch delivered the final calculations for the critical mass of uranium-235 metal to Robert Oppenheimer. It was Friday, April 13. He remembered he had completed the first fission experiment in Copenhagen on another Friday the 13th, in January, 1939.

Later that day, the metal blocks were taken back to the chemistry and metallurgy works where they were recast and shaped into pieces of metal to be used in the "Thin Man" bomb that Colonel Paul Tibbets and his B-29 crew were to drop over Hiroshima.

· 6 ·

Bainbridge held to his growing fear that America's own warplanes would destroy the first plutonium bomb, and with good reason.

The Trinity base site had seen rapid development; hundreds of military and scientific personnel prepared it for the coming event, building bunkers, laying miles of cable, erecting the steel tower, installing sophisticated equipment and instruments designed to reveal the secrets of an atomic explosion. To maintain rigid secrecy, the Trinity area—a tract of 470 square miles—had been declared a forbidden zone by the Army. It was patrolled by mounted soldiers and motorized guards, but that did not erase Bainbridge's fear of the airborne rookies.

On May 7, 1945, out on the flat desert, his team staged a unique trial explosion—using 100 tons of TNT. What made it different, apart from the size of the blast, was that 1,000 curies of radioactive waste from the plutonium-producing reactors was set among the huge pile of explosives, mounted nearly 30 feet above the ground. This was to test devices Herbert Anderson planned to use to detect the radioactive fingerprints of the atomic explosion. Unintentionally, it was a celebration firework display. It marked the end of the war against Nazi Germany. A few hours before the TNT was blown, German representatives signed the articles of unconditional surrender at General Eisenhower's headquarters in the French cathedral city of Rheims.

A week later, the Trinity base was again rocked by explosives—dropped by Air Force trainees. It happened on two separate nights; the planes droned over the forbidden zone and pilots and bombardiers mistook the Trinity lights for their illuminated target structures and let the missiles go. One bomb smashed into a barracks being used as a carpentry shop. Another hit stables, where a fire started, and panic ensued; fortunately, nobody was killed. Bainbridge also exploded; he asked Oppenheimer to press for radar-controlled artillery to drive the erring pilots away by firing warning smoke shells ahead of the advancing squadrons. It did not come about; and, until the end of the operation, the tower and the plutonium bomb stood on the mesa in constant danger from rookie bomb crews.

There were three other shadows hanging over Trinity. Oppenheimer had assigned a committee of six scientists to "ride herd" over the operation as a whole. They were at once dubbed the "Cowpunchers." They had a worrying time through May and June. The chemists and metallurgists had trouble with the first Hanford shipments of quantity lots of plutonium. Health hazards were a constant menace; the metal workers found it extremely difficult to handle since breathing a mere trace of the dust of element 94, with its extreme toxicity, exposed them to peril. There were unexpected, unusual aspects to the metal; in one state it just crumbled to gray dust. It had to be coated with a special, secret preparation once it was molded to the shapes of the bomb.

Troubles piled on delays. The day of the firing was set back on three occasions. Industry was behind in deliveries of key parts; firing circuits were two weeks behind promised supply; there were troubles with the detonators and worries about the state of the plutonium metal and the unknown factors of the accurate critical mass for plutonium. When the implosion method had been made firm, Edward Teller speculated that a strange phenomenon might occur, which could not be grasped easily in the state of knowledge then prevailing. It was "compression of the incompressible"; the transfer of the energy of the inward explosion could increase neutron effectiveness at the center of the severely crushed metal. It meant the size of the critical mass might be much smaller than was earlier thought.

On June 18, the first Los Alamos experts left for the Pacific island of Tinian; Colonel Paul Tibbets' 509 Group was there.

The "Cowpunchers" met on Sunday evening, June 24, 1945, to hear heartening news. In the Omega building, in Los Alamos Canyon, Frisch and his staff had built a plutonium assembly to a near point of explosion. Automatic controls had been invented to shut down the reaction at any point, and Frisch and his team had brought the mass to the indisputable point of imminent total fission. Among his team was a thin young Canadian, of Russian extraction, Louis Slotin, who was to head the assembly of the finished plutonium core of the first bomb—at the Trinity site—a brilliant young man who died from radiation in a laboratory accident after the war.

Frisch carried the news to the "Cowpunchers." The size of the critical mass of plutonium metal was established.

General Groves noted that President Truman and Prime Minister Churchill were meeting at Potsdam from July 15, and he wanted some news that would give Truman an ace in the hole in dealing with Stalin.

The convoy of Army sedans slipped out of Los Alamos early on Thursday, July 12, facing a blazing desert day. Armed men rode fore and aft, guarding the world's most costly metal, which sat in a specially designed carrying case on the back seat of one of the vehicles in the middle of the line.

Inside the box was the plutonium metal that would raise a man-made sun over the Jornada del Muerto four days later. If all went well with Trinity, a replica of this box and its contents would soon be on the way to the Pacific Island of Tinian for delivery above a Japanese city that would be selected, finally, by weather conditions. This trek 200 miles south of Los Alamos to the atomic base at Trinity was a rehearsal for the destruction of Nagasaki.

The metal in the box was ten days old. When it was new, it, too, was beautiful to behold—silvery, and immaculately shaped to the scientists' designs. But, just as with uranium-235 metal, it soon changed color as the surface reacted to air. With plutonium, however, surface corrosion was more threatening; not only would it endanger the precise tolerances needed to fit inside the implosion bomb, its toxic nature would imperil the men who had to handle it to put the bomb together in the desert. It took days for the chemists in Joe Kennedy's group to find the impervious coating. But the bomb was not yet complete; the outer shell of high explosives and detonators that would surround the metal inside the steel shell was even then being assembled in the Los

Alamos Canyon. It would make the journey the next day—Friday, July 13.

One other object was essential to the world's first atomic eruption. This was the initiator, a metal ball that would be set in the heart of the implosion device and which would, in a billionth of a second, fire neutrons that would trigger the 80 or so generations of fission chain reaction in the compressed plutonium. It rode in one of the other cars, along with a simulator, a matching ball of metal for assembly practice.

In the car, Dr. Philip Morrison held the two metal balls. He talked and joked with Louis Slotin and physicist Boyce McDaniel, who had worked on plutonium with Frisch's group.

Chuckling, Morrison played a "pea under the shell" game to pass time on the long, hot journey. He kept mixing the two balls and asking them to guess which was the right one. The true initiator was formed of an inner core of highly active alpha-emitting polonium and an outer shell of beryllium. McDaniel was terrified—the whole project was at stake if Morrison mixed the two balls of beryllium metal and the dummy one was inserted.

McDaniel had started the day with some concern; he had left his wife, Jane, anxious and worried. In the feverish action of the last days, many of the women in Los Alamos homes had become aware something big was in the air. He told her, "Please don't worry, honey. I'll only be gone a few days, and it's no more dangerous than riding with those WAC drivers on the roads for shopping in Santa Fe. It'll be O.K."

He was to remember that brave comparison in the next days.

Morrison finally told the secret of knowing the right ball. Hold them in your hand, he told McDaniel. One is warm, and one's cool. The warmer one is the real thing, with the active polonium inside shooting out energy to heat the beryllium cover. When these two metals were smashed together in the implosion, the neutron stream would flow.

They reached the assembly point at high noon. It was an old, four-room wooden ranch house that had belonged to a cattle-raising family named McDonald. Inside the wood-frame house, waiting in the searing heat, was Groves' deputy, Brigadier General Thomas Farrell. He signed an Army receipt that the bomb core had arrived and handed this to Louis Slotin. Then, in one of the wood-walled rooms, the scientists took off their shirts and put the pieces of metal together.

On the Friday morning, they drove very carefully the two miles to Point Zero, carrying the plutonium core, the initiator inserted in its bed.

A timber platform had been laid beneath the tower and a white tent erected over that to give shelter. The implosion device was inside—a five-foot diameter sphere with explosive plugs and detonators all pointing to the center where the core and initiator were to go. One section was missing, to allow access to the aperture in the metal where the bomb's heart was to fit. To achieve extreme density, the part had been machined with tolerances of thousandths of an inch. Great care had been taken at Los Alamos to ensure that *this* bomb core was the one made for *this* explosive shell.

Louis Slotin could not fit the part into the assembly. It jammed in the opening; it would not budge. He didn't try to force the section home, fearing to do damage. What had gone wrong? Had they brought two alien pieces by mistake? On top of this crisis, the weather laid another anxiety. A dry thunderstorm loomed overhead. McDaniel was nervous at the possibilities and went outside the tent. As he remembered, "It was a frightening situation. We were in trouble with the plutonium bomb under a 100-foot-high steel tower which protruded like a lightning rod above the desert, while the wind and lightning played around the structure."

He went back into the tent—and lo!—the problem had been solved—by nature. The metal assembly had been brought down from the cooler Los Alamos plateau, and the core and initiator had been in the desert all night. There was a difference in temperature, affecting size—time had resolved the problem, and the bomb was then "buttoned up."

There was one further task for McDaniel; a thin opening in the casing had been left for manganese wires he was to insert, reaching to the center of the device. Every four hours these would be replaced, and the ones removed he would take back to the shack, two miles away, to measure the induced radioactivity. He was checking the levels at the heart of the assembly, holding watch against any undue activity.

On Saturday morning, the tent was collapsed and the five-foot sphere was hauled by winch to a galvanized-iron shelter at the top. Every four hours thereafter, McDaniel would climb the steel ladder, in wind, driving rain, broiling sun—and when the sky was lit with lightning—renewing the manganese wires, taking the pulse of the first plutonium bomb.

Val Fitch was an enlisted soldier. Chance put him into the Special Engineer Detachment at Los Alamos and gave him the opportunity to be among the few hundred people to witness the first atomic explosion in history.

Fitch had spent a year at Los Alamos, picking up his own clues to the purpose of all the activity around him. From spring, 1945, he had a much clearer understanding, and he worked on the British physicist Titterton's three responsibilities in the desert test.

These were: to fit timing devices on the bomb to operate at the moment of firing; to provide timing markers for other groups for the milliseconds prior to eruption; and to send out the signal to trigger detonation at the exact moment. This last was to be done from the control bunker, six miles south of Point Zero, with automatic controls taking over seconds before implosion. Special timing devices were installed in an earthen bunker a half-mile from Point Zero.

The coaxial cable was hooked into the bomb when it was winched to the top of the tower; it was an arrangement that would only happen this once. In combat the job would be done by "Archie," a radar device which would close a relay signal system at the predetermined height above the target cities. This would be set in motion when the bombs fell from the B-29s; the system was started as wires were dragged from the weapons when leaving the bomb bays, and the firing system was primed in stages until, fifteen seconds later, a switch set to the altitude pressure closed and the bomb was exploded.

At Trinity, the same internal method was used, but the trigger would come through Titterton's cables and control devices. And there was trouble on that Saturday. Fitch climbed to the tower top for testing and found their cable circuit was broken. The group spent the rest of that day digging in the scorching heat of the desert for a half-mile to find where strain on a buried cable had broken the link. By dark on Saturday night the circuits were again correct.

The bomb was ready for firing; it had only to be finally armed by Bainbridge, using switches to which only he held the key.

Bainbridge spent that Sunday in the baking sun, solving a mountain of problems and annoyances. The top brass of the Manhattan Project were due in the afternoon—Groves, Bush, Conant, Lawrence, Arthur Compton, with Sir James Chadwick and Sir Geoffrey Taylor from England among them. Bainbridge stayed in the field. There was trouble with the short-wave radio frequency from the control bunker to the patrolling B-20 aircraft, which was supposed to be near the test site at the time of the explosion and then turn away to measure the kind of shock wave Colonel Tibbets would experience over Japan. The radio frequency needed to be different from that allotted to any local station; yet security had given them the Voice of America frequency. The guard cars and the dozens of jeeps, hooked on to the FM communication, had the same frequency as the rail freight yards at San Antonio, and on the atomic test bomb site they could hear shunting being ordered—and wondered whether the bomb orders were going into the rail yards. Bainbridge also discovered that the airport tower at Socorro could hear communications from the mesa at Trinity.

By supper time, his mood was tight. With Groves and the others at base camp, he listened to a colonel instructing how to avoid eye damage when the bomb went off, how to evacuate if there was a mishap. And there had been discussion, he learned, that there was a chance the bomb might set fire to the earth's atmosphere. Bainbridge was furious. He roundly condemned the talk as "thoughtless bravado."

It was wrong to talk like this, in front of soldiers ignorant of nuclear physics in view of the results of the special studies that had been done. Still very angry and knowing there was a long night and a day ahead, he retreated to his bunk at 8 P.M. and instructed a military policeman to wake him at 10 P.M.

By midnight, he was back at the foot of the tower. On orders from Groves, fearing last-minute sabotage, two military policemen shone torches every few minutes around the base of the steel frame. It was raining hard, and he knew the 2 A.M. scheduled shot would have to be canceled. He spent the next few hours there, in and out of the car, staring upward at the tower, speaking on a telephone fitted to one of the uprights, hoping the weather report would bring better news. At 2 A.M. a wet Boyce McDaniel made his last climb and regained his last manganese wire. Titterton and an assistant were then on their way to the control bunker from the instrument bunker, a half-mile away.

The rain had stopped—the clouds were flying. Just short of fifteen minutes to five in the morning, Bainbridge received the most favorable weather report; it was not ideal, but it was good enough to save waiting another day. He telephoned Groves and Oppenheimer. His last act at the tower was to turn on a set of ground lights for the pilot of the observation aircraft. The time of detonation was set at 5:30 A.M. He had several vehicles on hand for the run back to the bunker, just in case one or more broke down. He had bluntly told the drivers, that if a vehicle broke down, "Don't waste time waiting for help—get out fast! You can walk a mile or two in a half-hour, if you hurry."

He had minutes to get to control. There, he put the key in the arming switch—the last channel was opened to the explosion.

The time was 5:29 and 45 seconds. The timing sequence started, and they all went outside to lay down, to shade their faces and eyes from the glare of nuclear fire. With 45 seconds to go, the automatic timer took charge. Seconds before the explosive shell blew the plutonium segments into a "state of compressed incompressibility," the circuit designed by Ernest Titterton and his team sent out the pulses that would allow precise measurement by the waiting instruments.

The first atomic bomb was then beyond the hand of man.

· 7 ·

Otto Frisch left Los Alamos in the evening of Sunday, July 15, to drive south with colleagues to the viewing site Bainbridge had named Companga Hill. It was some 10 miles from Point Zero, and it was the only viewing site where Groves would allow anyone to stand in the open. As the long convoy of buses and cars thrust through the heavy rain, they knew the 2 A.M. scheduled shot would be delayed.

At Companga Hill, time dragged as they waited for news. Frisch went back to the car and tried to sleep for an hour or so; he felt as if he were "at an airport, held up by a defect."

Each time the loud speaker blared he sat up, hoping "the thing would go off." He had no deep feelings on what was to happen; what concerned him was that he had mislaid the welder's goggles he had been issued and, without them, he would have to be very careful that he was not blinded.

Another announcement—and another; then at last, the time was set for 5:30 A.M.

Back on the mesa at Los Alamos, hundreds of people had gathered for the flash in the sky that would signal triumph; many went home disappointed, thinking the scientists had failed. Only a few waited on till the edge of dawn.

On Companga Hill, Frisch knew his safest tactic would be to look away from Point Zero to the line of hills where the sky was still black. He recalled the occasion:

"Suddenly, the hills were bathed in a brilliant light; it was as though someone had turned the sun on with a switch. There was no sound. The light was similar to that of a camera flash. It seemed not to have color to my night-conditioned eyes. I tried to open my eyes wide to get used to the light. It surprised me; I had expected a brief flash, but this stayed for some seconds, and only then did it begin to dim. I was tired, but I told myself, 'Observe! You must observe; try to burn it into your memory so you can write it down later.' Then, when I thought it safe, I turned to see this pretty, perfect, red ball of fire about the size of the sun, connected to the ground by a short grey stem. . . . "

He sat on the ground, put his fingers in his ears and waited for the blast of the shock wave which came as an assault on the person, and then, long, low rumbling, like giant wagons running through the hills.

He did not talk to people. He went to sleep in the coach that took him back to Los Alamos, and there, he went to his room and wrote down all he had burned into his memory.

Bainbridge noted the plutonium bomb detonated fifteen seconds earlier than expected. Where he lay on a sheet of foam rubber, he could feel the heat on his neck, "disturbingly warm." Yet, there was a comfort in the fact, also exhilaration. He experienced wonderful relief—the duty he had feared he would not now have to face. If the bomb had fizzed, or hang-fired, it was he, as head of the test program, who would have had to walk alone to the steel tower to see what had gone wrong.

There was much more light than predicted. He could see Oscura Peak shining in the false sunrise, and he waited for it to dim before he took his first view of the explosion. Soon he could take off his goggles and see it with the naked eye.

"It was surrounded by a huge cloud of transparent purple air, produced in part by the radiations from the bomb and its fission-by-products. No one who saw it could forget it—a foul and awesome display."

He waited for the blast wave to batter past the control point, then he clambered to his feet to congratulate Oppenheimer on the success of the implosion bomb. And he finished by saying, "Now we are all sons of bitches!"

Oppenheimer understood. Later, he would tell the official observer, William Laurence, that, when the bomb exploded, lines from the Hindu Bhagavad-Gīta, came to his mind: "I am become Death, Shatterer of Worlds."

· 8 ·

Many hundreds watched the infernal incineration on the Jornada del Muerto; and impressions were varied. Among the observers were pioneers of uranium research; the man who discovered the neutron was there; so was the man who first used neutrons to bombard uranium and the man who did the first fission experiment. Also present was the scientist who was to be known as the "Father of the H-bomb," having then worked for three years toward a more terrible weapon in which this explosion would be merely the trigger. The man whose work led to the discovery of plutonium, Glenn Seaborg, was at home in Chicago, in bed.

All observers were stunned, some astounded, by the ferocity of the explosion. Yet, no human eye saw it all. The delicate covering of retinas would have been burned away in that first thousandth-of-a-second of the phantomic apparition of primordial fire. Only the fastest cameras, operating at a safe distance, captured the swift savagery of the birth of a nuclear fission explosion. In the first thousandth-of-a-second an ethereal form, without real shape, was there on the mesa, a quivering, ghostly, cosmic dome without definition—a blinding, shimmering, fierce light, more brilliant than any that had shone on earth; and in less than a hundredth-of-a-second it was a convulsed, boiling immensity, lifting off the desert floor as the head of an enormous toadstool, a loathsome growth with a coiling, frilly skirt of dancing sand being scorched to glass, a nuclear fungus suppurating into the air, swelling and lifting, outward and upward. And between three-hundredths and five-hundredths of a second, the unearthly fire was convoluted by the raging release of the ancient energy of a billion-billion-billion atoms created from the uranium that had fed the giant reactors; like the maddened gases of the sun, the lines of atomic force rolled over and around and down and inward. And the head of the gross, fiery toadstool lifted—in one second from fission it was ascending, dragging the dust of all it had liquidated, and all it had liberated, into a stalk of lethal radiation. Up, up, it rose, into the young morning air, higher than Everest, touching 40,000 feet, with searchlights tracking its journey eastward for the first eighteen miles, to be camouflaged among the clouds, to circle the globe, spreading, again and again, and to lay its trail of dust for future rains to bring back to earth, to deliver into the grass-animal-man food chain the noxious products of the world's first atomic fire—the shattered fragments of fission.

The explosive energy of the Trinity bomb was the equivalent of 18,600 tons of TNT. This force, the mightiest ever achieved by man, came from only part of melon-shaped segments of plutonium; heat and pressure in the first fission generations at the solid heart of the metal destroyed the rest before it had time to react.

Both the intensity of light and the explosive power were greater than predicted, so great that the military men at once declared World War II ended. The phenomena revealed by instruments, however, was more indicative than sheer blast and destruction; here was the empirical evidence of the deadliness of by-products which Chadwick had forecast in his assessment of the weapon for M.A.U.D., some four years earlier.

Radioactivity! Eight neutron detectors were set around Point Zero; all but two were vaporized, and those two showed neutrons

200 meters from the blast to have the alarming energy of 300 million electron volts. *Here was the shape of things to come in a future neutron deterrent weapon.*

A great deal of data was gathered on the character of high-energy fission; Robert Wilson set, close to the bomb, an ionization chamber—a refinement of the tool with which Frisch proved fission in 1939. The chamber was connected to the instrument bunker by a coaxial cable. In the first millisecond of searing heat, Wilson's ionization chamber was dust, but the brief measurement signals raced at the speed of light to the bunker, just ahead of the vaporizing cable. Other tests were overwhelmed by the titanic outburst of radiation: the flooding gamma rays and X rays which turned the air blue and fogged the films of many camera experiments. The most simple test was by Enrico Fermi. He held a handful of scraps of paper and, as the shock wave came charging like a wall of air, he let them go—and so intent was he on seeing where they landed, so he could calculate the power of the bomb, he remembered nothing of seeing the fireball in its most dramatic moments or hearing the sound of the explosion. For the third time, Fermi missed observing the action of fission.

Herbert Anderson made a sortie into the area of Point Zero in a lead-shielded tank, driven by Sergeant Bill Smith, later that morning after the exultant generals and scientific mandarins had returned to Washington, Los Alamos, and other points. Without leaving the protection of the lead and steel, Anderson operated the arm to scoop a sample of soil from the crater, which was six feet deep and 1,200 feet in diameter; they could see the fission blast had turned the sand to a vivid green color. Smith then turned the tank, and they rumbled to safety.

From five miles away Bainbridge sighed with relief; it had been planned to use two vehicles, in case one broke down; and the second tank was out of order that morning.

Anderson returned to Los Alamos with the fission fragments impregnated in the soil by the immense explosion. With Oppenheimer hovering anxiously while he worked to establish the radioactive ratios of fission by-products, Anderson found isotopes of known substances, from elements 36 to 60. There was a wide range of breakages from the cores of 92 and 94, among them isotopes of cerium, caesium, strontium, and iodine and the gases xenon and krypton. In the parade of their presence at Trinity, the United States had a measuring tool to sample the world's air to detect whether any other nation challenged American atomic supremacy. Indeed, Anderson laid the base for the stunning event in September, 1949, when a patrolling B-29 found the radioactive fingerprints of the atmospheric explosion of the first Soviet uranium bomb.

Val Fitch and a fellow soldier-technician in Titterton's electronics group drove a panel truck into the desert, to the instrument bunker, a half-mile west of the devastated crater. The spot where they had worked on Saturday in the searing heat was unrecognizable. The soil that had banked the bunker had been blasted away, leaving the bare concrete box standing in the sun. The cables they had toiled to bury beneath the surface had been uprooted, scorched, frayed and had been thrown back over the bunker by the blast. Fission had savagely rejected man's efforts to conceal and protect.

The films they recovered from the bunker were useless; great waves of gamma rays and X rays had penetrated to them—just as radiation from uranium salts had penetrated to Becquerel's photo-plate long before. It didn't matter; had the bomb *not* fired, they would have served as a valuable diagnostic tool.

A day or two later, when Fitch and a civilian technician, Calvin Linton, were leaving the control bunker to return to Los Alamos, they made a more daring penetration. They passed the bunker where they had been on July 18; then they went closer to Point Zero, to the edge of the violated desert where the green sand was glasslike; and, with no shielding lead or steel, Fitch leaned precariously from the open truck door and scooped a sample of fission-impregnated soil, to be his personal memento. He recalled, "Not knowing the radiation level precisely, we hastily retreated from the area."

· 9 ·

The Trinity triumph brought no relaxation in the work at Los Alamos; indeed, it added concern. Years before, these scientists had set out on the trail of this weapon of destruction—now proved to be much more fearsome than expected—and the drive had been the fear the Third Reich would be ahead with atomic weapons. Now Groves' agents had revealed a puny, piddling German effort. The exiled scientists, whose anxiety in the early war years had urged America on to the uranium stage, now had doubts of unannounced atomic war on the people of Japan. Leo Szilard again enlisted Einstein to gain an interview with the President, but the letter had been found, unopened, at Hyde Park, after Roosevelt's death. Widely supported, signed declarations had appeared—one with many prominent names, including that of Seaborg, was sent to the White House but was not opened until ten days after the destruction of Hiroshima.

In the metal workshops on the "Hill," on the mesa of New Mexico, there was little inhibition; the driving force there was speed. Supporting the use of atomic arms, by what was now the world's most powerful warring nation, were other fears, like those that governed the thoughts of the Secretary of War, Henry Stimson. He saw the prospect of a physical invasion of Japan, in the light of suicide pilots and fanatical troops dying for the honor of defending their native land, "an appalling cost in the precious lives of American servicemen". The military planners estimated that at least one million men would die invading Japan. There was this other way to end the bloody war.

The second major production of plutonium metal was shaped and delivered for the combat weapon by July 23—one week after Trinity. On the following day, the critical shape of uranium-235 metal was delivered; this was to become a fissioning mass when struck by the projectile slug of 235 metal fired by the gun assembly when the bomb was some 1,800 feet above the port of Hiroshima.

So eagerly had this metal been prepared, it was delivered for combat use 48 hours ahead of schedule. It made no difference. When the silver bird of doom, which its captain had named *Enola Gay*, delivered its lethal cargo—at 9:15 A.M. (Tinian time) on August 6, 1945—weather in Japan had delayed, by four days, the moment when the awful blinding light filled the sky.

It was nearly 42 years after the night when Ernest Rutherford stood in the garden in Paris and saw the same light shining from the Curies' first radium sample, when Pierre Curie spoke his fears of this power falling into the hands of "warring men."

In the Washington office of the Manhattan Project, on Pennsylvania Avenue, Flora L. Little—of Jackson, Mississippi—was having trouble with her teleprinter. Fretful at delay, conscious of the long list of important messages she had to send for General Groves, she tapped out her irritation and frustration on the keyboard.

"It isnt yr machine.Its mine.And there isnt a thing can be done.The repair man has been here all day . . . and I have loads to go to u tonight . . . we'll have to do it this way.A few lines at a time.OK?"

Some 1,500 miles to the west, in the Los Alamos office of the Director, Dr. Robert Oppenheimer, Mildred Weiss of New Orleans, Lousiana, sent back her O.K.

The message from Groves was in five parts, tapped Flora Little; then she started, using "PD" for period:

FLASHED FROM PLANE FIVE MINUTES AFTER RELEASE . . . QUOTE CLEAR CUT RESULTS COMMA IN ALL RESPECTS SUCCESSFUL PD EXCEEDED TEST IN VISIBLE EFFECTS PD NORMAL CONDITIONS OBTAINED IN AIRCRAFT AFTER DELIVERY WAS ACCOMPLISHED PD VISUAL ATTACK ON HIROSHIMA. . . . WITH ONLY ONE-TENTH CLOUD COVER PD FLACK AND FIGHTERS ABSENT UNQUOTE.

AFTER RETURN TO BASE FARRELL SENT FOLLOWING . . . QUOTE A LARGE OPENING IN CLOUD COVER DIRECTLY OVER TARGET MADE BOMBING FAVORABLE PD EXCELLENT RECORD REPORTED . . . NO APPRECIABLE NOTICE OF SOUND PD BRIGHT DAYLIGHT CAUSED FLASH TO BE LESS BLINDING THAN TRINITY PD BALL OF FIRE CHANGED IN A FEW SECONDS TO PURPLE CLOUDS AND BOILING AND UPWARD SWIRLING FLAMES . . . RISE OF WHITE CLOUD FASTER THAN AT TRINITY PD IT WAS ONE THIRD GREATER IN DIAMETER REACHING THIRTY THOUSAND FEET IN THREE MINUTES PD MAXIMUM ALTITUDE AT LEAST FORTY THOUSAND FEET WITH FLATTENED TOP PD COMBAT PLANE STILL OBSERVED CLOUD AT THREE HUNDRED AND SIXTY THREE MILES AWAY. . . .

The news ran like a flash fire through Los Alamos.

In his laboratory in Los Alamos Canyon, Frisch heard a commotion—feet pounding along the corridor, then a voice shouting, "Hey! Listen, everybody. Listen . . . they've destroyed Hiroshima!"

So that was the target! There had been guarded talk among the members of the Co-ordinating Council of what cities would qualify. Frisch went out into the corridor where there was a throng of scientists, research assistants, army technicians, all abuzz with the news. Frisch later told this writer that he heard a loud voice he didn't recognize: "A great success, they say. Bigger than Trinity, and the whole city annihilated. The early estimate is at least 100,000 Japs have been wiped out!"

There were cheers, and someone yelled, "That's great news, after all the work. Let's celebrate!"

And another voice added, "Yeah, let's ring the hotel and book a big table for tonight. We'll all get good and drunk!"

Frisch wanted terribly to be alone. He had always claimed—and did so again, many years later—to be "essentially a rational human being"; but now the tight rein of control sagged. He turned away from the babbling group and went into the laboratory where he had done the experiments on the critical masses of uranium-235 and element-94. The knife was excruciatingly painful—"at least 100,000 Japs . . . wiped out"; ". . . let's celebrate!"

He recalled he had not experienced such a depth of emotion since the morning of Friday, January 13, 1939, when he had made the first uranium fission experiment in Copenhagen and the janitor brought the cable with the news his father had been released.

He sat alone in the laboratory and thought how, "after all, this is war. It's been grim, awful, and maybe this catastrophe will end all the killing and hate, and bring peace to the world. Maybe in that sense it will do good."

It had not been wrong to build the bomb because the Germans—even the Japanese—might well have done the same and would have no compunction about its use had the atomic dice rolled their way. He reasoned that he had no cause to regret what he had done, the part he had played. In all conscience he could say that originally *he did not want chain reaction to work!* In 1939, when Niels Bohr argued there were too many inert uranium-238 atoms for the chain reaction to give an explosive weapon, he had felt gladness. It had wiped away the specter that this terrible force would come into the hands of men. But when it became clear from his work with Peierls, and with Simon and others, that the specter still hung in the shadows, he had felt a duty to all the oppressed and the menaced peoples.

Yes! On the basic fact of working to save the free world from the Thousand-Year Reich, it was justified, essentially rational.

He did not join the celebrations that night. He sat alone in his room, and his mind roamed back to the single critical decision that led him to all this, to the time when he decided not to follow his mother as a concert pianist, but to try to emulate his aunt, Lise Meitner.

· 10 ·

Not all the dead from the atomic bomb project were counted in Hiroshima and Nagasaki. The lethal power lurking in all uranium atoms claimed its first American victims in the very heart of the complex that produced the weapons.

Frisch's assistant in the perilous critical assembly work was Harry Krikor Daghlian, who knew well, from the start of his days at Los Alamos, that he lived with the world's most dangerous substance. He had been present in the Omega laboratory in Los Alamos Canyon when Frisch came within two seconds of death.

The bombs destined for Japan had been delivered; twelve days after the destruction of Nagasaki, on the evening of Tuesday, August 21, Daghlian went back to his bench in Omega to work on an idea for a critical assembly of plutonium, with a tamper of neutron-reflecting metal, aimed at reducing the size of the combat warhead.

In the final steps of building, his fingers slipped on a block of fissile metal—the assembly went supercritical, heat and blue light were under his body, the neutrons were swamping his torso. In a desperate lunge he threw the material across the floor—too late. At five minutes to ten in the evening of that day, Harry Daghlian sustained a painless dose of radiation lasting a few seconds. He knew he would die but held out against molecular deterioration of his tissues for 25 days. His mother, a sister, and a brother were brought to Los Alamos to be with him; he finally succumbed on September 15. On the next day, his mother was presented with a check, in compensation, for $10,000.

There was to be a repetition of the tragedy. Precisely nine months to the day after Daghlian's accident Dr. Louis Slotin was also engaged on critical assembly experiments. With Slotin, it was the slip of a screwdriver which allowed two hemispheres of uranium-235 metal to combine long enough to saturate his body with neutrons and gamma radiation. Slotin was not alone in the laboratory to which the critical assembly work had been transferred; there were six other people in the room. They escaped death. The radiation dose was fiercer in his case. His agony lasted for nine days, and his parents, sister, and brother were with him when he died. His mother also received a compensatory check for $10,000.

These deaths added to the mounting knowledge of the anatomical effects of critical amounts of radiation from fission. The official record noted, "Perhaps the most spectacular pictures concerning the health group were the complete photographic coverage, in black-and-white, and in color, of the effects of the tragic radiation accidents which resulted in Harry Daghlian's death in September, 1945, and Louis Slotin's death in May, 1946."

· 11 ·

With Hiroshima and Nagasaki incinerated, with the world in awe of this new, mighty striking power, with fissile material flowing through Hanford and Oak Ridge into the secret bomb laboratories, Groves could now tell his countrymen something of the immense achievement and the feverish drive instilled into the greatest organization of science and industry and military muscle in history.

A steady supply of atomic bombs was assured. America was invincible, the sole atomic power on the planet. It was time to be generous, to admit a debt. In his office in Washington, Groves took pride in pinning the Medal of Merit—America's highest civil award—on Edgar Sengier. The uranium from the Shinkolobwe mine, said the general, had been crucial to the atomic bomb undertaking; in all their dealings, Sengier had behaved as a wise and true patriot, a man of freedom, a statesman, and ambassador for his country.

Not to be outdone, the British recognized the vital nature of the Shinkolobwe ores in the ending of World War II; Segnier was knighted, the only British knight to hold the American Medal of Merit.

Groves took recognition of the Congo uranium further; on the day Japan officially surrendered, he escorted Sir Edgar Sengier to the White House to meet President Harry Truman, and said,

"Mr. President, this is the man without whose help and ready assistance the Manhattan Project would have been a failure."

On that same day, they had a celebration lunch. At the Army Navy Club many officers congratulated both Groves and Sengier. The astute Belgian tycoon went on record as muttering to one officer, "Mais aux États-Unis plus qu'ailleurs est éphemère."

It was a comment afterward freely translated into nuclear terms: "Fame in the United States has a short half-life."

The words were sadly prophetic.

In 1960, when bloody rebellion first swept the Congo, Sengier was disgusted by the attitude of the United States, by lack of action in the United Nations, by the carnage unleashed among the black people. He sent back the Medal of Merit to the White House.

For the man whose observation made the world's richest deposit of uranium available to launch America into the atomic era—in war and in peace—there was no such recognition. Major Robert R. Sharp, D.S.O., M.C., French Croix de Guerre (with palms), had farmed his fields in Rhodesia through the years of World War II. Not until the atomic bomb was exploded over Japan did Sharp know of the consequences of the day he kicked the rock on the top of the African kopje to find the world's richest uranium mine.

Six years after Hiroshima, he was awarded one of hundreds of similar honors—the Order of the British Empire, the O.B.E.

TRANSMUTATION

The peoples of the world ached for an end to the violence and yearned for a future free from the hate and destruction which had circled the globe. They were ready for the promise of a new age. Twenty months after the atom bomb fell on Japan, the impact still reverberated around the planet; scenes of instant carnage and devastation were vivid in the public mind, in no way dispelled by arguments that this was merely an added dimension to the old problem of war. True, there had been the horror of death from the sky in Rotterdam, Hamburg, Leningrad, Coventry, Dresden, Tokyo and London—the latter still the only major city to know both suborbital rocket and robot-bomb assault. True, a man was as dead from an arrow as from atomic flash. It was not the numbers of the dead that branded Hiroshima and Nagasaki on human awareness, nor was it the sudden, swift catastrophe—a single bomb from an aircraft; these were refinements of the long, mad history of human conflict. There was something else that burst upon the world with the fission explosions of uranium and plutonium.

Winston Churchill had labeled this force "supernatural." It presented the unknown, it delivered the unfelt, invisible death stroke: silent killer-rays, death-dealing neutrons, swift solar heat, lingering radioactivity. All were alien to human experience.

Yet this frightening new force could be used as a mighty tool for rebuilding—a weapon against hunger, poverty, and sickness. And in the United States, the first civil agency to control atomic energy matters was established, with David E. Lilienthal, the driving former chairman of the Tennessee Valley Authority, at its head.

On the night of Saturday, April 19, 1947, he strode to the rostrum in the big ballroom of the Statler Hotel, in Washington, to face an audience of the most powerful and influential men in the land—government members, judges, senators, congressional leaders, soldiers, scientists, top journalists. This was the annual banquet of the American Society of Newspaper Editors.

Lilienthal held a small cylinder of silvery metal above his head—no bigger than a cigarette pack—as he proclaimed, "This substance is now the central object of discussion in the councils of the world. This is uranium.

"In this small cylinder is power we can release, and control, that equals the burning of 2,500 tons of coal. Probably the most important decisions we will ever make in our history as a nation concerns the use we will make of uranium, of the course and development of atomic energy. . . ."

There was personal triumph for Lilienthal. Control of the gargantuan atomic empire General Groves built had been wrested by President Truman from the hands of the Army, but not without a bitter factional fight. Lilienthal had been attacked, vilified by Senate opponents, with the same insinuations of Communist association that were to destroy Robert Oppenheimer. He had emerged, his appointment confirmed, and now he was ready to take atomic energy into the new peace, to prepare the ground for a future of cheap, limitless energy. Ahead were refinements of Fermi's crude graphite-uranium pile, of the big Hanford plutonium producers into stations producing steam from nuclear fire to turn the electricity-generating turbines: new, clean, silent palaces of power! Energy to burn—to change seawater to fresh, to irrigate thirsting fields, to create fertilizers to feed the hungry crops; reactors to make diagnostic isotopes to trace the processes of life in man, animals, and plants, to detect disease and to heal: a mighty civil construction tool to cut new canals, blow instant harbors, gouge great reservoirs—all with a single stroke. The plan was to put uranium into overalls to work for man.

Lilienthal told his audience, "These are not just technical issues. They are human matters. They should be discussed in the open, in the public forum, not treated as secrets. So I'm proposing a broad and sustained program of education at the grass roots of every community in the land—and this is the task of the media."

His audience rose to applaud. And, through the throng came the most popular figure in the gathering—conqueror of Nazi

Germany, to be elected president in 1952. General of the Army Dwight D. Eisenhower pumped Lilienthal's hand and in his brisk manner declared, "I'm on your team!"

There was a fatal flaw in Lilienthal's shining dream; like the metal he held aloft, the golden age he envisioned would tarnish with exposure. The new civil body he led also had responsibility for the development of atomic weaponry; and even as he spoke, a Soviet team led by the brilliant Igor Kurchatov—a team that included Peter Kapitsa—had already achieved a controlled chain reaction in a graphite and uranium assembly as Fermi had done in Chicago in December, 1942.

In the rare, cold atmosphere, some 18,000 feet above Russia's Kamchatka Peninsula, the patrolling weather-sampling bomber droned from Japan to Alaska. There were special apertures in the wings of the plane; inside were filters and adhering surface devices similar to those that Herbert Anderson had first used for the Trinity test shot.

The next day was Saturday, September 3, 1949, the tenth anniversary of the beginning of World War II. In the late afternoon, teleprinters began to rap out their alarming message for American leaders. From radioactivity gathered above the Bering Strait, it was known that Russia had detonated an atomic bomb in the atmosphere. There was a second nuclear power in the world; the daydream of American atomic monopoly was shattered.

The fissile material the Russians used was plutonium—a further shock.

Andrei Vishinsky, standing at the United Nations rostrum a few weeks later, described a new, shining age similar to that of David Lilienthal. He said the Soviet Union planned a future generation of uranium-burning stations with fuel enriched with the isotope 235, with heavy water to moderate the neutrons, and would use explosives "to move unwanted mountains, alter the course of great rivers, irrigate deserts and make them bloom." Despite this, Soviet nuclear capacity was more than a challenge— it was a threat. It propelled the development of the device that would be known as the "Hell Bomb."

Fermi and Teller had debated the prospect over lunch eight years before. Borrow the technique of the stars, that was the idea!

If the uranium fission device worked and the projected heat of more than 20 million degrees centigrade—believed necessary to operate the fusion, or sticking together, of hydrogen atoms in the heart of the sun—could be achieved, then, weight for weight, fusion of hydrogen isotopes* would liberate ten times the energy released by uranium fission. There was a limit to the size of the critical mass of fission material; there would be no limit to the size of this "super-bomb"—the hydrogen bomb.

The devices that would be detonators for this fusion weapon had first to be attained. And then, with the war over, Teller's fierce conviction that the "super" would one day be needed was borne out when the B-29 captured fission fingerprints over the Bering Strait.

Decision-making on the H-bomb was a process fraught with apprehension and soul-searching. Among the world's most eminent scientists fears and doubts were spreading. Einstein was saying, "to avoid this menace has become the most urgent problem of our days." Oppenheimer was appalled, with colleagues, that this "weapon of genocide, an intolerable threat to the future of the human race," should be considered; so staggering in its power it would be wasted against armies in the field and could only be used to exterminate the larger cities of the world.

Fermi and others, in a declaration, saw the "super" taking mankind into "realms of great catastrophe"—that it should be renounced by the pledge of all nations on earth. The British physicist Blackett—with whom Frisch had worked in prewar days—found classical allusion in the lines of Ariosto, who had his hero Orlando declare,

> O! Curs'd device. Base implement of death!
> Fram'd in the black Tartarean realms beneath!
> By Beelzebub's malicious art designed,
> To ruin all the race of human kind!

The father of plutonium, Seaborg, wrote Oppenheimer that he would have to hear very strong argument before he "could take on enough courage to recommend that the H-bomb should *not* be built." Seaborg was then on his way to receive the Nobel

---

* These were deuterium and tritium (which Oliphant discovered in 1934). The existence of both isotopes was forecast by Rutherford, along with the neutron, in 1920.

Prize for chemistry with his Berkeley colleague, Ed McMillan; together, they stood in Stockholm where, almost a half-century earlier, Pierre Curie had warned of these supernatural forces falling into the wrong hands.

On the last day of January, 1950, President Truman issued his decision: "It is part of my responsibility as Commander-in-Chief of the Armed Forces to see to it that our country is able to defend itself against any possible aggressor. Accordingly, I have directed the Atomic Energy Commission to continue its work on all forms of atomic weapons, including the so-called hydrogen or super-bomb. Like all other work in the field of atomic weapons, it is being, and will be, carried forward on a basis consistent with the overall objective of our program for peace, and security. . . ."

Truman's decision soon had strategic validity. At midyear, Communist troops of North Korea invaded the south and the world lurched toward a terrible new war, with General Mac-Arthur urging the use of atomic bombs to discourage the Chinese. The Soviets were waging the Cold War; Berlin had been blockaded in 1948 and 1949, and the United States had replied with the Berlin airlift; and within weeks of Truman's decision came disclosures of espionage that shocked politicians and public alike and drove a wedge between the victorious Allies. Hopes that had lingered for what the official history calls "a tightly integrated program with the British and Canadians" died with these revelations.

Espionage had first raised an ugly prospect when Allan Nunn May, who had worked on the British M.A.U.D. project as early as 1940, confessed in 1946 to having approached a Soviet diplomat in Ottawa while engaged on the British-Canadian atomic work. He had then been impelled by his association with the Communist Party early in life to pass a sample of uranium isotope extract to the Russian "ally"; he was sentenced to ten years in prison. It was another four years, however, before the extent of the Soviet network was revealed with the arrest of Klaus Fuchs by Scotland Yard officers, in February, 1950.

A morose, humorless man, Fuchs had fled Germany in 1932. He worked in Birmingham with Peierls on the M.A.U.D. project. When the U.K. team went to America he was included and thus had first-hand knowledge of gaseous diffusion; he was one of few scientists with exact information on the construction of the atom bomb. A few days after President Truman's statement that work would start on the H-bomb, Fuchs was arrested in London and, heading for a sentence of 14 years at the Old Bailey, told detectives of his many acts of treachery.

In America, the Fuchs trail led federal agents to Harry Gold, a chemist from Philadelphia, and his confession to spying for Russia showed a link with Fuchs and agents in the Soviet diplomatic corps. Gold's arrest in late May, 1950, led to David Greenglass. A former Army sergeant who had worked at Los Alamos, Greenglass admitted passing secret details to his sister and brother-in-law, Ethel and Julius Rosenberg. The harrowing Rosenberg trial and their subsequent execution caused wide controversy; and the fear of proliferation of nuclear weapons penetrated deeper into daily life.

Then, in October, 1952, the third member of the atomic club appeared. At Montebello, off the west coast of Australia, the British exploded their first bomb, a plutonium device; and they were on their way to hydrogen weapons. Also, there was deep concern in Washington over the French situation; the now avowed Communist, Frédéric Joliot-Curie, and his wife Irène, held top executive positions in the French atomic energy commission. All this added to perturbation and served to drive forward the work on the "super."

Some 5,000 miles off America's West Coast, in the Marshall Islands, was the chain of small atolls known as Eniwetok. The little islands ran from Sanil to Elugelab, Teiter, Bogairikk, and Bogon. On the last day of October, 1952, one of those islands was wiped from the face of the earth in one second.

A green-white super-sun, hundreds of times brighter than mid-day, rose over the atoll; the dazzling eruption made the blue Pacific diamond bright. It was the most savage explosion achieved by man to that time; it made the Trinity and Japanese outbursts seem puny. The explosive force was equal to 5 million tons of TNT. Had it exploded over any metropolis, it would have caused utter devastation over an area seven miles wide. It created a stupendous fireball, more than three miles in diameter, which rose quickly to the heavens, touching 40,000 feet in two minutes and going higher. Where Elugelab had been, the ocean poured its waters into a crater more than a mile across, and deep enough to hold fourteen buildings the size of the Pentagon.

They called this explosion "Mike." The fission trigger was a new design, more powerful than any before, and it fused atoms of hydrogen isotopes, as does the sun, to liberate energy. It was not only the greatest explosion achieved by man—it was the foulest injection into the atmosphere of toxic by-products of shattered uranium atoms. The great cloud-head pierced the stratosphere. It had assaulted the earth; now it assailed the outer limit of the planet's biosphere. In the cold, swift winds of that rare altitude, some 25 miles above the ocean, the poison cloud spread, widened, expanded, to lace the thin atmosphere with a cargo of unnatural radioactivity that can be created only in such an explosion.

In the pale sunlight of the Siberian north, Peter Kapitsa stood looking across Lake Tumuldy toward the structure holding the new bomb. To his right ran the tundra, out to where the Yenisey River snaked toward Tomsk, the capital of Siberia, a thousand kilometers southward. This bleak, remote place had been inhabited for millenia by Russia's nomadic aborigines, small tribes of primitive people beyond the fringe of civilization. For weeks, Soviet military police scoured this wide region, driving the tribesmen and their families into holding areas, like wild cattle. Huge lumbering air transports carried the bewildered nomads away to safety; some 150 aircraft were needed. Now, the preparations were done; it was Wednesday, August 12, 1953, and Kapitsa stood patiently, waiting for the moment when his years of calculations and experiment would meet confirmation—or failure.

It had been easy for him to follow the progress of research in the West; the Soviet network was superb, and every word of every published article came to his desk from the world's scientific journals. In 1943, he knew of Los Alamos, its purpose and problems. He had known of Oppenheimer's needs, materials he sought, how much he was getting; even how the secret work progressed from month to month. While the Germans hammered toward Moscow, as they fought the titanic battle of Stalingrad, Kapitsa and his colleagues followed paths that led from public notices of fission in the Western journals; and neptunium's birth guided them to transuranic work. Thus, Roosevelt's ultrasecrecy toward an ally fighting for its existence did no more than seed suspicion and distrust. At the time America's H-bomb burst above Eniwetok, Kapitsa and his colleagues were well ahead on their own model, which was now ready for test-firing on little Popey Island, in Lake Tumuldy.

This test shot would reveal his executive ability as chief of the team. The bomb was Kapitsa's main creation.

The colossal fireball outstripped his expectation. The enormous fulminating cloud of green-blue-orange fire left him aghast, as it had many scientists; it created a severe anxiety that was to cause his downfall in the Soviet scientific hierarchy. He turned away when the cloud rose high above his head; he was pale, and still sweating. He called to his dog, "Come, Borgia—come!"

Days later, when it was possible to enter the area of scorched earth, searchers found charred bodies of six nomads who had escaped the Soviet dragnet to become the first victims of a Russian hydrogen bomb. The world was not told about them. In the following March, an awesome U.S. explosion in the Pacific—with the energy of 14-million tons of TNT—dropped fall-out on a little Japanese fishing boat near Bikini. The name of the vessel was splashed in newspaper headlines; it was called the *Lucky Dragon*. Some one hundred miles from the scene, the crew suffered injury to their health from fallout.

More than any other event, it was the *Lucky Dragon* accident, the impact of high-energy fallout on the few Japanese fishermen, that sparked public outcry.

The cloud fired over Lake Tumuldy soon spread across Siberia. Within days, the American detection network had samples of the by-products under study in Los Alamos. Not only was this a hydrogen bomb, exploded a mere 9 months after "Mike," its fuel contained an element not used in the Marshall Islands tests. The Russian scientists had taken a simpler path by using lithium in place of tritium, lithium being cheaper and more easily accessible than tritium. It was an advance which American scientists copied. It shook to the roots the notion the Soviets depended on spies and traitors for their nuclear capability.

Secretly, Peter Kapitsa wrote letters to friends in the West, urging every effort by scientists to campaign for a world moratorium on atomic weapons. Carried by a trusted colleague attending an international congress in South America, the messages fell into the hands of the Soviet secret police, and Kapitsa was demoted, disgraced, and left in a position of no eminence or consequence. Visitors to Russia thereafter, such as his former associate at the Cavendish, Professor Oliphant, found him withdrawn about his work and "not very forthcoming."

Five days before Christmas, 1951, final preparations were made for an historic experiment in a squat, white concrete building, on the flat Idaho desert, that housed the most complex atomic reactor yet built. This was the creation of Walter Zinn, director of the Argonne Laboratory, and pioneer of the Fermi team in Chicago.

Into the core of this Experimental Breeder Reactor No. 1 (ERB-1) Zinn's team had loaded 52 kilograms of uranium-235—small rods which he classified as "standard lean," "standard fat," and "long fat." With these, Zinn achieved an even distribution of neutrons from the fissioning uranium. A blanket of inert uranium-238 surrounded the core of 235; neutrons from the rods would smash into the 238 and breed plutonium to be used in other reactors. It was planned to produce more plutonium than the amount of 235 used. It was a prototype that came to be likened to a coal-burning station where, for each shovel of coal thrown into the furnace, two were taken out the other side. But that was not all! On this day, when Zinn took the reactor to the point of critical fission, there was another purpose. The heat rose in the core and was carried by a sodium-potassium mixture through pipes, pumps, to a heat exchanger, and there it made steam and spun the turbines. The generator hummed; inside and outside the concrete building the lights gleamed into the dark, wintry afternoon: Electricity from fission!

Zinn made an entry in the station logbook: "1:23—Load dissipator connected to generator. Electricity flows from atomic energy. Rough estimate indicates 45 kilowatts."

Man had won useful power from the uranium atom. The path was opened to mammoth stations across the world silently churning out tens of millions of watts of electrical energy. Only four years later, on July 17, 1955, another reactor in Idaho was supplying all the electricity needed by the small town of Arco. A year before that—with a burst of publicity rare for Soviet atomic undertakings—the world heard of a nuclear power station operating at Obninsk, near Moscow, with an output of 5,000 kilowatts. The Soviet model was a water-cooled graphite reactor, using enriched uranium; it was very similar to a reactor Zinn had designed to power the world's first nuclear submarine, *Nautilus* (shades of Jules Verne!), which circled the world without refueling. Zinn's work was claimed, later, to have laid the foundation for nuclear power "in America, and across the world."

In May, 1956, British scientists completed their first nuclear-power reactor—Calder Hall. They planned to use the plutonium (other than what was needed for weapons) to fuel their first breeder reactor at Dounreay, in Scotland.

Two other nations, France and Canada, were also on the doorstep of nuclear-power production. Canada, with the advantage of her own uranium supply and the wartime work done in that country by British and French scientists, had opened a successful era of reactors that were refinements of Joliot-Curie's original "atomic balloon." They used natural uranium, with heavy water to slow the neutrons for greater release of energy.

France had been quick off the mark. Lew Kowarski, and other wartime exiles who had worked in Canada, joined Joliot-Curie who, with Raoul Dautry, had talked with General de Gaulle soon after liberation. The work dropped in 1940 was resumed. The French scientists, however, no longer had the great source of raw materials from Union Minière; not even Belgium had access to that. Under the intergovernment agreement, which Groves piloted in 1944, all ores extracted from Shinkolobwe went to America—for ten years after the war. Luck was with Joliot-Curie; a rail car of concentrate of uranium from Oolen had passed through Toulon and was located in Morocco; another carload sent to Le Havre, laden with yellow cake, was found untouched. Locals had believed it to be some kind of dyestuff. They had a small haul of 12 tons. It was enough to start a heavy-water-moderated assembly. And the Norwegians, who had put their heavy water plant back into production, stood fast by the agreement made with Lieutenant Allier in March, 1940. All the work done up to the time of the German occupation now bore fruit—for France.

They did not make metal from their uranium; they used oxide, as was planned in the original "balloon," and it worked wonderfully well. On February 15, 1948, the first chain-reacting pile in Europe went critical, at Châtillon, outside Paris. They were heady days for the French team; Joliot-Curie was chief of the atomic energy commission and, with his wife, in high standing among their countrymen. They called the first reactor *ZOE,* a zero energy experimental pile.

By the end of 1949 ZOE had produced their first plutonium, and a big atomic energy center was building at Saclay; in addition, Geiger counters had uncovered a large field of uranium near Limoges. Yet Joliot-Curie and Irène's loyalty to wartime association with Communists threatened French relationships with powerful allies and menaced security. Joliot-Curie was a victim of his own public utterances, and de Gaulle dismissed him from office on April 27, 1950. His wife was also barred from access to national secrets. The noted chemist Bertram Goldschmidt, who had worked with Seaborg, said of his colleague and countryman, "One can only regret the growing hold of politics on the greatest French nuclear physicist ... it was to separate him more and more from atomic energy, in which he had played such a vital role for our country."

The dismissal of Irène and Joliot-Curie proved to be a curtain raiser to events in America and also a classic symptom of fear bred from the notion of uranium's potency in the hands of the Soviets. Robert Oppenheimer was to suffer the same swift cut: obliteration from all scenes of America's atomic energy projects. Because of their political associations—freely admitted—the past services of all three to their nations were spurned and they were grievously hurt in their minds and in their public careers.

For Robert Oppenheimer the blow was most savage and sudden. In the postwar years he served as spokesman of the powerful American scientific elite that had called down cosmic power to end the war. Symbolic of the new age of atomic energy, he was given the nation's highest civil award, the Medal of Merit—which Sengier too had received—and honors showered on him from many nations. His prestige at its height, he was asked in 1953 by the British Broadcasting Corporation to deliver their noted annual Reith Lectures—and when he went to the microphone in London some foresight seemed to touch his brilliant mind as he said, "The power to change is not always good. As new instruments of newly massive terror add to the ferocity and totality of war ... man's ever-present preoccupation with improving his lot—alleviating hunger, poverty, and exploitation—

must be brought into harmony with the need to ... resort to organized violence between nations.

"The increasingly expert destruction of man's spirit by the power of the police—more wicked ... more awful than the ravages of nature's own hand—is another such power, good only if never to be used."

Even as his voice went on the air in Britain an axe was being sharpened in Washington that would fall on his public service—when the time became ripe.

The shock of the Russian H-bomb development still reverberated through Washington corridors; Senator Joseph McCarthy was belaboring the Atomic Energy Commission for laxity in security; and before long a letter written by a senior official was on the desk of President Eisenhower with the covert opinion that Oppenheimer was probably a "Soviet agent in disguise."

Lewis Strauss—a war-time admiral—was then heading the AEC; he saw Eisenhower, and "a blank wall" between Oppenheimer and official data was at once thrown up. Strauss handed the former chief of Los Alamos a dreadful Christmas present. On the day before Christmas Eve, 1953, the general manager of the AEC, General Nichols—the same Nichols who first signed an agreement with Sengier for the Shinkolobwe uranium—sent a letter to Oppenheimer: ". . . there has developed a considerable question whether your access to A.E.C. work will endanger the common defense . . . ."

The old associations renounced in 1940 to Compton, details of his past life, and his opposition (among a phalanx of physicists) to development of the hydrogen weapons were all to be used in dismantling his character as "defective with falsehood, evasion, misrepresentation."

He could accept dismissal as a consultant, he was told, and nothing more would be said; or he could fight the allegations in a private court of inquiry.

Oppenheimer wrote that such an attitude would mean accepting "the view that I am not fit to serve this government that I have now served for twelve years. This I cannot do. . . ."

His whole life was laid bare at the hearing, supposed to be *in camera*, but afterward leaked to the media. Nothing he said, or scientific associates said, altered the predestined path. Though he was idolized by the public and admired for his intelligence and culture, his sad, handsome eyes gazed from the pages of news-papers in the spring of 1954.

His public demise coincided with the upsurge of fear roused by the fallout from the Bikini H-bomb on the Japanese fishermen in the *Lucky Dragon*. Oppenheimer retired from the spotlight to an academic role at the Institute of Advanced Studies in Princeton, to serve his last years in teaching. Uproar, protest flowed from many directions. The noted commentators, Joseph and Stewart Alsop, accused the AEC of pillorying a man who had served his country well and with no evidence of disloyalty against him other than his opposition to the "Hell Bomb." He had performed a task of scientific coordination which, John J. McCloy had said, "could have been done by no other American"; Oppenheimer's recompense, said the Alsop brothers, was a "miscarriage of jus-tice."

His dismissal—by four votes to one—sent shock waves through the nation, already plagued by allegations of un-American ac-tivities. Not until a new Administration was installed in the White House and at the AEC in the early 1960s was Oppenhei-mer officially exonerated. President Kennedy invited him to a White House dinner for Nobel laureates and planned to present him with the AEC's top recognition, the Fermi Award.

The presentation, because of Dallas, fell to President Johnson to perform late in 1963, shortly before Oppenheimer's death, with the words, ". . . you are here to receive formal recognition for your many contributions to the advancement of science in our nation. Your leadership in the development of an outstanding school of theoretical physics in the United States and your contri-bution to our basic knowledge make your achievements unique in the scientific world."

Handing the citation to the President at that ceremony was the new chairman of the AEC—Glenn Seaborg.

The lines of atomic energy development by mid-1958 con-verged toward inevitable controversy. The stage was set for the great atomic dilemma.

Man's knowledge of the structure of matter was growing; new and mysterious materials were appearing on the scene. Funda-mental components of the universe were under astonishing ma-nipulation; radiochemists went beyond cosmic creation into the rare fields of "transurania." In the superbly equipped Met-Lab in Chicago, Seaborg and his co-workers trod deep into Rutherford's "no man's land." They made, and isolated, two new transuranics beyond plutonium and named them—but only after careful thought, for there were no more known planets beyond Pluto to be invoked. Such was the hectic seeking, the pinning down of these elusive substances, that they were for a time known as "delirium" and "pandemonium." When their existence was proved, they were named: "americium," for element 95, and "curium," for element 96. In peacetime, the trail was blazed deeper; new man-made elements were added, elements 97 through to 103—also carefully named: berkelium, californium, einsteinium, fermium, mendelevium, nobelium, and lawrencium. Awkward names for such rare, exquisite substances—and no place for Rutherford, or Bohr, or Klaproth.

In that same year of 1958 the growth of the technology of nuclear-power generation led to a thousand uses for man-made isotopes—in medicine, industry, and agriculture; there was strengthening of the hope that this immense new force would be a rich gift applied to the well-being of mankind and that this would make atomic energy more acceptable. But there were also bombs by the hundreds, and along with them the overhanging dread of fallout.

Two decades after its contents had razed Hiroshima and Nagasaki, the great uranium mine in Katanga was no more. Through the years, the Shinkolobwe mine yielded more than 40,000 tons of the richest uranium ore.

For the Union Minière company, with the great mine dead, all that was left of the halcyon days of world monopoly when radium fetched $3 million an ounce, was a few packets of the element—some 200 grams in all—which, today, radiate in the dark of secret bunkers, one near Niagara Falls, one in America's Midwest, and one near Oolen, where even the old extraction plant has disappeared. The radium boom was over from the moment uranium-burning reactors could drench* cobalt and other substances in the neutron maelstrom of fission. The element's place as a penetrative tool was won by cheaper, easily

* The cobalt is made radioactive in reactors.

produced materials; its therapeutic value was usurped by ranks of isotopes made to order.

In the critical decades of its life, when it was the mainstay of American atomic enterprise, Shinkolobwe seemed a vast Eldorado. Then, when its life was done, when the legions of prospectors and geologists went out with their Geiger counters, it was disclosed as a geologic freak only by its remarkable richness. Uranium was not rare—it was, in fact, as common as copper. Assays showed it to be in most rocks, in the folds of ancient hills, in pre-Cambrian structures, in the seas; one survey concluded that every square mile of soil, to a foot deep, in America's Midwest contained some three tons of uranium, enough 235 for a critical mass.

The pace of the search for uranium, the speed of bringing new deposits into production has never been equaled with any other mineral; veins of richly laced deposits—a few as old as Shinkolobwe—ancient sediments, disseminated uranium types in crystalline rocks were all pursued with ardor. By one decade from the end of the war the discovery of deposits had proliferated—there were finds in Utah, at White Canyon, Marysvale, Moab, San Rafael; in South Africa; and in Rum Jungle, Australia; at Todilto and Laguna in New Mexico; at Monument, Arizona; in the Black Hills of South Dakota; in Canada's Beaverlodge and Saskatchewan. Thorium was also found in quantity in India, Ceylon, Brazil, and in the United States. There was no shortage of radioactive raw material; yet it had to be concentrated enough to make mining worthwhile to feed the ranks of an estimated 200 nuclear power reactors in 50 lands across the globe. It had to satisfy the maw of the weapons materials centers, building even then toward frightening capacity.

By 1960, annual production of uranium in the non-Communist world was about 30,000 tons, most of it earmarked for weapons. With demands for power reactors, operating and planned, the annual need in 1985 is expected to reach 100,000 tons, and to double every five years thereafter, to well above one million tons annually by the year 2001. Presently known economic reserves are put at little more than one million tons; with expected growth, mining experts see formidable supply problems for the nuclear power industry, unless exploration is advanced, all available material mined, and background reserves lifted to 4 million tons by the end of the century.

To this enormous release of energy from natural and enriched uranium will be added the plutonium resources. Only rough estimates are possible, due to the vagaries of public opinion and the opposing desires and needs for light, warmth, energy; but some forecasters can see a half-million kilograms of plutonium being produced, handled, and re-used annually in the Western world, by 1999.

Only guesses are possible on the Soviet Union's true access to fissile material. In 1945, when, by Big Three power play, General Eisenhower's troops had to retreat to demarcated lines, those early deposits that Klaproth and Marie Curie had investigated in Saxony and Bohemia came under Soviet control. Forced labor was used, and the ores were quickly plundered; and so, a curt, crisp order went out from the Kremlin to geologists in the newly enslaved lands—to Poland, Hungary, Rumania, East Germany—"Find uranium!"

Indications have leaked to the West of intensive scouring of the Soviet's own extensive lands, producing huge finds—more than 300,000 tons in Tadzhikstan, Kirghistan, Tyuya-Mutun, Taboshar, and the Fergana Valley. Certainly, no special bonuses were paid, and mining was not held up by protesting demonstrators and union action. Uranium is used in Russia without quibble—for power, for civil engineering, for weapons.

Projected use of uranium fission for electrical power generation in the Western world alone is mind-boggling—a thousand nuclear reactors by the year 2000. It also presents a staggering and bewildering problem of high-energy waste products with long, active lives.

Ubiquitous uranium is transformed in man's fission furnaces into a threat to the total environment, a threat realized in every corner of the world and constantly fanned by committed campaigners and political tacticians. But not without good reasons, reasons powerful enough to persuade people that science and technology, which have made such giant leaps in the last few decades, will not be able to deal with this threat to the human biosphere. The fear of radioactive waste leaks, the near-impossible "melt-down" of reactors, has grown from the madness of weaponry displays in the decades after the war. In retrospect, it was an awful cavalcade.

Today, with five members in the nuclear arms club, some 500

great explosions have been fired above ground, unleashing a total force equal to 500 million tons of TNT. Soviet bombs contributed some 300 million tons to this atmospheric assault. In those early days military weaponeers referred to these monsters as "dirty" and "cleaner" bombs; and when Britain, and then France and China, joined the hunt for the ultimate weapon, colossal power became, in a word, multi-megaton. The most stupendous explosions were triggered as high as 200,000 feet above the face of the earth. Carried aloft by rockets, these were holocausts of terrifying force, each capable of expunging the greatest city and millions of people from the planet. They were testimony to a new phase of sophisticated nuclear technology; management of critical masses of uranium-235, and its daughter, plutonium, set amid chilled, solidified hydrogen-type fuels, touched new peaks of horror.

Each battering assault on the earth's air had more serious implication; even the so-called cleaner bombs laced the upper atmosphere with complex creations of fission and fusion, substances of long and very active life that came to be known simply as "fallout."

These horrific injections into the world's envelope of air held a legacy for every man, woman, child, every vertebrate creature on the ground. Foul clouds carrying radiostrontium, iodine, caesium, and other fission by-products enfolded the biosphere—even the frozen poles—and each shower of rain or snow brought down its cargo of radiation, to mingle with the soil and the water of the earth, to enter the food chain, for the processes of biology to make selection on where they would be finally deposited in the body—the bones, glands, thyroid, the kidneys. Not just once, but constantly, they filter through this chain; and these toxins of the mad race for the ultimate horror weapon will continue coming down for decades, in decreasing amounts—while detached scientific spokesmen pronounce on whether or not they are "significant" to the population as a whole.

The spokesmen discussed, as a tactic, whether or not these toxins exceeded the background radiation from rocks, cosmic rays, the sun, or medical X rays; but, most importantly, fallout was feared and resented and roused a furor around the globe, with experts fearing irreparable genetic damage. Einstein warned humanity of "self-annihilation"; and in his 1955 Christmas message, Pope Pius was moved to plead with world leaders for an end to nuclear pollution of the world's atmosphere.

The leaders were all fully alive to the menace. President Eisenhower made a dramatic bid for the "age of the peaceful atom" and told the United Nations that "the United States pledges before the world its determination to help solve the fearful atomic dilemma . . . to find the way by which the miraculous inventiveness of man shall not be dedicated to death, but consecrated to his life."

There was little hope the words would bear fruit. In May, 1960, the U-2 aircraft piloted by Gary Powers was shot down over Russia; in the next year America severed diplomatic relations with Cuba; and this was followed in April by the Bay of Pigs debacle. In August, 1961, the border with West Germany was closed and work started on the Berlin Wall, intended to stop people fleeing to the West. Russian technical demonstration was vivid that year. Yuri Gargarin became the first human to orbit earth.

On October 23 and again on October 30 the Soviets defied worldwide protests. The two largest man-made explosions in history were fired into the atmosphere. The first equaled 25 million tons of TNT; the second had the power of 50 megatons. The stupendous power of the explosions sent shock waves around the world. The "miraculous inventiveness of man" seemed to many to be directly aimed at the death of humanity.

These decades of apocalyptic explosions not only laced the air with foul fission fragments—they sowed the earth with seeds of rage that grew into violent protest. Once people saw the unborn in danger, the horror became psychotic. For the plain men and women of the earth, all forms of atomic energy were born in original sin with the bombs. Out of that has come the nuclear dilemma and the acids of suspicion that transmuted the gold of uranium's promise to the dross of fear.

The promise of Lilienthal's shining future, Eisenhower's hopes for benefit from uranium, are now barricaded behind this wall of fear; nuclear power development, uranium exploitation is hamstrung by claims of possible massive poisoning, of great nuclear disaster, all prophesied by extreme activists—at the very point in history when man can see an ending to the supply of fossil fuel.

The American physicist Ralph Lapp has calculated that by 2000 A.D. some 200,000 Americans will die from cancer triggered by *natural* radiation in the environment—but, that, allowing for 1,000 reactors by the end of the century, the routine release of fission toxins will account for no more than 90 deaths. To replace such nuclear-power resources, he estimates, some two billion tons of coal would need to burn, with unaccountable health effects and impact on the world's atmospheric structure—and coal mining has already cost at least 100,000 lives in America in mine accidents and coal transportation.

Does this defend uranium against psychotic fear?

Even nuclear physicists have fears; among them are some of the still-living pioneers on the remarkable journey of intellectual exploration. One of those who helped erect imperishable landmarks in human discovery is Emeritus Professor Sir Mark Oliphant, who ended his public career recently as governor of South Australia, once one of Rutherford's young team:

> I worked on the secret atom bomb project from the start. I travelled to the United States more than 20 times during the war, taking the secrets of radar to them, helping to set new uranium work in motion, working with Ernest Lawrence on magnetic separation. For a long time I was a fervent advocate of the peaceful uses of atomic energy. Then, I saw the threat. I added my name to a manifesto with Albert Einstein and Bertrand Russell, warning the world against the dangers of fallout from atmospheric bomb tests. The Canadian-born American millionaire, Cyrus Eaton, put his aircraft at our disposal for a meeting in his home town, Pugwash, and from that the Pugwash Congress emerged. It was at once labelled Communist by the people who refused to see the dangers. I was denied a visa to visit America; nobody called me a Communist, but they said I had given the Reds bullets to fire. It did not change my opinion, my apprehension. The existence of nuclear weaponry is the biggest challenge facing mankind today. Until this threat is eliminated, nobody can be certain of peace or that they will see another day dawn on earth. Nothing now can stop prolifera-

tion of atomic weapons and atomic power across the earth. Nothing can wipe away the poison uranium has brought to mankind's affairs.

Dr. Glenn Seaborg, the man who brought plutonium into the world, who for ten years from 1961 was chairman of the U.S. Atomic Energy Commission, also sees danger; but he sees the problems as a challenge which holds out promise of great benefit to mankind:

> The discovery of uranium fission truly opened new horizons ... for us to expand our search for truth, both inward to the heart of matter, and outward to the far reaches of the universe. On the question of plutonium: it is true to say, as with all the power of science and technology at our command today, that what will come of plutonium depends on how mankind chooses to use it. Perhaps fear of the massive destruction it could bring will be the deciding factor to bring men together to choose reason rather than conflict as a means to settle differences. Perhaps, also, it will be the power in plutonium that will be used constructively and beneficially to help men achieve the things essential to world stability—a more widely shared abundance, and a lasting peace. We stand on the threshold of intensive development of plutonium. I am confident that in the next decades plutonium will bear the fruit of present promise and that new and important contributions to mankind will be discovered.
>
> Perhaps it would be better if uranium and plutonium did not exist; but, that is not a viable option. Therefore, given the fact that these things do exist, that uranium and plutonium cannot be swept back under the carpet of knowledge, the most sensible solution is to learn how to make use of them. All nations, to differing degree, have an energy problem which will grow more acute, so we need to solve the difficulties of managing these materials, of containing their poisonous nature. I think this can be managed. The potential for war or peace, for devastation of the planet or the lifting of all men to new standards of living, brings into sharpest focus the major dilemma of our times. ...

Can man, who now holds his destiny in his own hands,

act with enough wisdom, patience, and understanding to choose the right path? I think he can. I think history will lead us all to a better and brighter tomorrow.

History holds a different prospect for Otto Robert Frisch, former director of high-energy physics at the Cavendish, Jacksonian Professor of Natural Philosophy at Cambridge—and disillusioned with the glowing promise of a nuclear age:

> Uranium-burning stations, plutonium breeders, even reactive lithium in hydrogen power stations will always be dangerous to human beings. I look forward to a world where men no longer depend on fossil fuels—not on coal, wood, oil—nor uranium and hydrogen. It may take a long time, but man must break the bad habit of using whatever happens to be lying around to meet his growing energy needs. In the end, he will be compelled to stop his wandering into such thickets of danger and to turn back to the original source of all energy. He must turn to the sun. I am sure that, finally, there will be dramatic advance in development of techniques for storing and using this natural font of power.

Pray that Otto Robert Frisch is right, once again: that there will appear on earth one day soon another triumvirate to match the genius of Rutherford, Einstein, and Bohr, to unravel the complex mystery of how to store solar power for man's direct use.

Until then, beyond the remaining fossil fuels, the uranium series offers ten thousand times the power humanity has yet used; and human hunger for energy will not be denied. Mankind will still wander into Frisch's "thickets of danger," and uranium and plutonium will not be swept back under Seaborg's "carpet of knowledge." Until the sun has been harnessed, mankind will depend on the family of the deadly element, and—hopefully— learn how to avoid its perils.

A P P E N D I X

INDEX OF PRINCIPAL SCIENTISTS

ABELSON, Philip: University of California graduate who was close to uranium fission in 1938, cooperated in discovery of element 93, neptunium. Worked on separation of U-235 for American bomb project.

ALLIER, Jacques: Secret agent with Banque de France who snatched the sole source of heavy water from the Germans in Norway in 1940. Special courier to British uranium committee in April, 1940.

ANDERSON, Herbert L.: Columbia University graduate; worked on neutrons and built a cyclotron; took part in first American fission experiment with Dunning; cooperated with Fermi on first uranium reactor at Chicago. Fashioned fallout tests at Trinity explosion, later used to detect Soviet bombs.

ASTON, F.W.: British disciple of Rutherford at the Cavendish; invented the mass spectrograph for first precise measurement of isotopes. Worked in 1931 on gaseous separation of light isotopes.

BAINBRIDGE, Kenneth: American physics student at the Cavendish, in 1935 where he heard Rutherford first speak of "chain reaction" notion. Attended M.A.U.D. meeting to debate feasibility of uranium bomb. Worked on implosion device for plutonium at Los Alamos; put in overall charge of Trinity. Later, professor of physics at Harvard.

BECQUEREL, Henri Antoine: third-generation physicist at natural history museum, Paris. Intrigued by X ray discovery, found uranium could expose photographic plates and so triggered discovery of radioactivity; research by Rutherford and Thomson in Cambridge and Marie and Pierre Curie in Paris stemmed from his work.

BETHE, Hans: German-born physicist who worked in U.S. from 1935 and led a theoretical group at Los Alamos. First worked out the fusion process to explain the sun's energy production.

BLACKETT, Patrick (Lord): British physicist with Rutherford; former naval officer and a pioneer of cloud chamber photography; associated with Frisch; later became president of the Royal Society.

BOHR, Niels Henrik: Student of Rutherford in 1911; added understanding of electron shell to planetary atom and first linked it with quantum theory; devised notion of "water-droplet" nucleus. In 1939 proposed U-235 for the key role in uranium fission. Great humanitarian. Played special role in problem solving at Los Alamos. Deeply concerned at atomic threat, but failed to impress Western leaders with his plans for a peaceful introduction of nuclear energy.

BOTHE, Walther: With fellow German physicist, H. Becker, in 1930 bombarded boron and beryllium with alpha radiation and reported a powerful emission that could penetrate 10 inches of lead. They wrongly identified this as gamma rays; through the interest of the Joliot-Curies it led Chadwick in Cambridge to the neutron in 1932. Bothe joined with Heisenberg to work on Germany's abortive war-time uranium investigation.

BRIGGS, Lyman J.: Chairman of President Roosevelt's uranium committee in 1939. A former agricultural scientist, he was criticized for hesitancy and delay.

BUSH, Vannevar: President of Carnegie Institute, made head of U.S. Office of Scientific Research and Development, and given overall charge of the uranium problem. Close adviser to Roosevelt and chief collaborator in Manhattan Project to General Leslie Groves.

CHADWICK, James: Rutherford associate at Manchester and Cambridge; interned in Germany in 1914–18 while working with Geiger. Crowned 12-year search in 1932 with discovery of neutron. First advised U.K. government on release in fission of dangerous radioactivity; took part in development of A-bomb as member of M.A.U.D. Committee. Led British team to work in Los Alamos.

COCKROFT, John: Member of team at Cavendish, 1932, first to split the atom, with accelerated particles. Member of the "Kapitsa" group and when professor of physics at Cambridge, 1939, joined the research team of the Ministry of Supply; became member of the M.A.U.D. committee to work on uranium problem. In 1945 was appointed director of the British Atomic Energy Establishment, then headed the Atomic Energy Authority.

CONANT, James: President of Harvard, close associate of Vannevar Bush, became his deputy in OSRD and headed America's national defense research project. Played a prominent role in pressing for atomic bomb from 1942 onward.

CURIE, Marie (née Sklodowska): Studied under Becquerel at Sorbonne and chose emanation from uranium for doctoral thesis. Married Pierre in 1895; together, they discovered and named radioactivity and went on to isolate radium and polonium; awarded joint Nobel Prize with Becquerel in 1903. Pierre died in street accident in 1906; Marie won second Nobel Prize, 1911.

JOLIOT-CURIE, Frédéric and Irène: Married 1926, were co-discoverers of artificial radioactivity, 1934. Their observations led to discovery of neutron. Irène first reported lanthanum in uranium after neutron bombardment, 1938; this led to Berlin experiments later elucidated by Frisch and Meitner. Frédéric sketched first draft of uranium test reactor, planned to build uranium bomb in Sahara; with colleagues, discovered chain reaction and perceived role of 235 isotope and the critical mass of uranium. Lodged first patents for reactor development, 1939; failed to escape from France with heavy water and remained for the duration working with the underground. Became head of the French Commissariat for Atomic Energy (CEA); was sacked by de Gaulle for Communist affiliations.

DAGHLIAN, Harry K.: Member of critical assembly group at Los Alamos who died September, 1945, from radiation caused by critical mass almost fissioning during experiment.

DARWIN, Charles Galton: Was present in Manchester in 1911 when Rutherford revealed concept of the planetary atom. Represented U.K. in America and was involved in Manhattan Project arrangements. Grandson of the evolutionist.

DEMPSTER, Arthur: Canadian-born physicist; used mass spectrograph to seek isotopes in all elements in Chicago. In 1935 discovered uranium 235, the isotope that is key to uranium fission.

DUNNING, John: Professor of Physics, Columbia University; studied neutrons, took part in first American experiment to fission uranium, January, 1939.

EINSTEIN, Albert: Theoretical genius whose work gave meaning to the puzzling radiations discovered by Rutherford and the Curies. General relativity theory, with the well-known equation, proposed the equivalence of mass and energy for the first time, in 1905. In 1939, signed letter urging President Roosevelt to order research into uranium potential and sources of supply.

FERMI, Enrico: First to bombard uranium with neutrons but failed to read results correctly, missing fission discovery. Went to U.S. after Nobel Prize award, 1938; worked on slow neutron studies and later planned and supervised erection of world's first uranium reactor, at Chicago University, December 1942. Discussed with Edward Teller the physical pathway to creating hydrogen explosions.

FOWLER, Ralph, H.: Rutherford's son-in-law. Discovery of neutron announced in his room at Cambridge. Served Britain in U.S. and Canada during Manhattan Project; first urged American study into plutonium possibilities.

FRISCH, Otto Robert: Born Vienna, studied in Germany and Britain, then joined Bohr in Copenhagen. Became expert on neutrons; discussed the puzzling Hahn-Strassmann results of uranium bombardment with his aunt, Lise Meitner, in Sweden, Christmas, 1938. Afterward performed first physical experiment on the uranium neutron disintegration to which he gave the name of "fission". In wartime Britain, with Rudolf Peierls, first saw possibility of an exploding critical mass of uranium isotope 235; made physical measurements to convince British authorities of uranium bomb possibility; at Los Alamos made critical measurements on the size of the plutonium and uranium metal cores for first A-bombs. Later, became director of the Cavendish, Cambridge.

GEIGER, Hans: Rutherford protégé who gave his name to the counter for detecting atomic particles; first developed in Montreal, refined later in Germany.

GENTNER, Wolfgang: German physicist, worked with Joliot-Curie in Paris before and during the war.

GROVES, Leslie R.: U.S. Army general who performed administrative feats from September, 1942, to produce first atomic combat he dispersed funds of $2 billion to build huge reactors to produce plutonium and magnetic separation plants for uranium-235 supplies.

GOLDSCHMIDT, Bertram: French chemist, worked under Marie Curie and later with Seaborg in Chicago. Studied plutonium as a base for British extraction plants and became chief of the French atomic energy commission.

HAHN, Otto: German chemist sometimes regarded as "father of radiochemistry"; worked in London and with Rutherford in Montreal in 1905, discoverer of new aspects of radioactivity and several elements. Developed sensitive method of measuring the recoil of atoms. Met Lise Meitner in 1912 to start 30 years of collaboration in which they jointly discovered protactinium. In 1934 repeated Fermi's experiments seeking transuranic substances. In December, 1938, repeated Irène Joliot-Curie's experiment of neutron bombardment of uranium and found barium by-product, a result which was elucidated as fission by Frisch and Meitner.

HALBAN, Hans von: Jewish scientist, born Vienna, worked with Frisch in Copenhagen, and at Collège de France with the Joliot-Curie team in Paris. Escaped with heavy water when France collapsed, 1940, and in England with Kowarski showed overwhelming proof of their discovery of chain reaction in fissioning uranium.

HEISENBERG, Werner: German physicist, studied with Bohr, made notable contributions to understanding of nucleus; in charge of German uranium work at the Kaiser Wilhelm Institute, in Berlin, during war.

HEVÉSY, Georg von: Born in Budapest, worked both with Rutherford and Bohr, and later cooperated with Frisch; took part in the Manhattan Project. Discoverer of element 72, hafnium, and originated the use of isotopes in tracing processes in biology, for which he won the Nobel Prize in 1943.

KAPITSA, Pyotor (Peter): Son of a tzarist general, joined Rutherford at Cavendish in late 1920s making impression with his flair for generation of high currents used in particle accelerators. On holiday in Moscow was prevented from leaving Russia; known to have worked on first Soviet uranium and plutonium bombs and to have led the building of Russia's first hydrogen bomb.

KENNEDY, Joseph: Colleague of Seaborg at Berkeley, in discovery and development of plutonium, 1940–41. Led work on fissile metal fabrication at Los Alamos bomb laboratories.

KLAPROTH, Martin Heinrich: Born in Weringerode, became pioneer chemist and original discoverer of element 92 which he named uranium, in 1789, in tribute to discovery of planet Uranus; also discovered beryllium, tellurium, chrome, cerium, and zirconium.

KOWARSKI, Lew: Born in Russia, escaped Bolshevik revolution as a boy, and came to Paris; worked with Perrin and joined Joliot-Curie as assistant and secretary. In January, 1939, witnessed fission experiment and took part in establishing feasibility of chain reaction in uranium. Escaped to Britain with Halban, and constructed and operated first experimental heavy-water reactor. Joined with British team on work in Canada; after the war joined French atomic energy effort and then went to Geneva to join CERN—the Centre Européen de la Recherche Nucléaire.

LANGEVIN, Paul: Prominent French physicist, studied with Rutherford at Cambridge, close associate of Pierre and Marie Curie. French patriot, head of the French Academy, was persecuted by the Gestapo and fled to Switzerland for the duration.

LAWRENCE, Ernest Orlando: American genius of the uranium story. Invented the cyclotron (Nobel Prize, 1939), which produced the first plutonium. Became driving force for urgent action in American bomb development in 1941. Devised magnetic "race tracks" in cooperation with Oliphant to produce separated uranium isotopes for use in bomb building at Los Alamos.

LINDEMANN, Frederick (later Lord Cherwell): Adviser to Winston Churchill on science; first cast doubt on probability of uranium bombs, later urged the building of atomic weapons in Britain and argued against passing secrets freely to America.

McMILLAN, Edwin: Early in 1940 working with Abelson in Berkeley, California, discovered first transuranic element—No. 93—and named it neptunium. Later joined secret radar work at M.I.T. and joined Oppenheimer at Los Alamos in 1943 to work as radiochemist on bomb problems.

MENDELEEV, Dimitri Ivanovitch: Originator of the famed periodic table, which first gave elements their places in an ascending scale of atomic weight; son of a blind Siberian school teacher.

MEITNER, Lise: Co-discoverer of uranium fission with nephew Otto Frisch; co-discoverer of element 91 with Otto Hahn. Fled from Nazi racial laws from Berlin to Sweden in 1938, and later that year solved the mystery of uranium nucleus disintegration. Later became a citizen of Sweden.

MOON, Philip: Associate of Oliphant in Birminghan 1939 on radar, and then became a member of the M.A.U.D. Committee.

MOSELEY, Henry G. J.: Rutherford disciple at Manchester, evolved the "atomic number" to equal the number of charges in the nucleus and thus caused rearrangement of part of the periodic table; killed at Gallipoli, 1915.

NODDACK, Ida: Hungarian physicist and chemist, discovered element 75, rhenium; in 1935 made first suggestion that the Fermi results of neutron bombardment of uranium could be due to the nuclei breaking apart. Her words were unheeded.

OLIPHANT, Mark: Australian-born physicist, made first attempt to transmute uranium, 1925. Under Rutherford at the Cavendish discovered tritium (3-part hydrogen) in 1934. Was chief associate of Rutherford, then went to Birmingham to work on secret radar research; became prominent member of M.A.U.D. Committee. Gave new impetus to U.S. uranium work, September, 1941, on one of twenty visits made there during war. In the Manhattan Project worked with his friend Ernest Lawrence on magnetic separation of U-235. After the war became advocate of abolition of nuclear weapons. Became Governor of South Australia.

OPPENHEIMER, Robert: Famed American scientist who headed the physical development of the first atomic bombs. Studied in major European centers, then headed physics studies at University of California, Berkeley. As head of Los Alamos earned the title "father of the atom bomb." Argued against development of hydrogen weapons; earlier association with known Communists led to his dismissal from the U.S. AEC by President Eisenhower.

PEIERLS, Rudolf: Fled Nazi racial decrees, 1933, for a post at Birmingham and there later worked with Frisch on use of uranium-235 as an explosive. Suggested gaseous diffusion for separation of isotope and achieved mechanical method for creating critical mass in a bomb delivered by aircraft. After Chadwick, led the British scientific team at Los Alamos.

PERRIN, Francis: Theoretical member of Joliot-Curie team which framed the first four patent applications for nuclear power production. First worked on the idea of a critical mass for uranium explosion. Later, became head of the French Atomic Energy Commission.

PERLMAN, Isadore: Seaborg's colleague at Berkeley and Chicago; in Manhattan Project cooperated on plutonium creation and extraction. Became professor of physics at Israel's Weizmann Institute.

PLACZEK, Georg: Studied with Rutherford and Bohr; in Copenhagen worked on neutrons with Frisch, later escaped German occupation of Denmark and took part in the Manhattan Project.

ROENTGEN, Wilhelm: German physicist who discovered X rays, 1895, findings which led to discovery of radiation from uranium by Becquerel.

ROSENFELD, Leon: Assistant to Niels Bohr who inadvertently broke the news of fission to American physicists, January, 1939.

RUTHERFORD, Ernest (Lord): New Zealand-born giant of atomic discovery. Began and ended career of 40 years at Cavendish Laboratory, Cambridge. Discovered first radiated particles, named them alpha, beta, 1896. With Soddy, in 1902, discovered spontaneous disintegration of atoms, in Montreal. In Manchester, 1911, revealed structure of the planetary atom. In subsequent years trained most leading figures in the uranium epic. Achieved first true transmutation in 1919; predicted existence of a neutral particle—neutron—and two isotopes of hydrogen, deuterium and tritium. Led work on development of devices of detection on discovery of neutron and on particle accelerators which finally "split the atom" in 1932. Died 1937.

SEABORG, Glenn Theodore: Radiochemist at Berkeley, worked with McMillan on transuranic elements and led team which discovered element 94, plutonium. Directed wartime plutonium work in Chicago, devised extraction procedures for Manhattan Project. Was chairman of U.S. Atomic Energy Commission for 10 years from 1960.

SEGRÉ, Emilio: Colleague of Fermi in Rome neutron bombardment, 1934, later worked in America. Discovered element 43, technetium, in an irradiated sample of uranium provided by Ernest Lawrence. With Seaborg, discovered fission capability of plutonium. After the war worked on transuranic elements and won Nobel Prize, 1959, for creating anti-proton.

SENGIER, Edgar: Chief of Belgian mining company, Union Miniére, which developed the Shinkolobwe mine in Katanga. Left Brussels for New York when war started; shipped direct to New York uranium ore without which America's atomic bomb project would not have been successful.

SHARP, Robert R.: British prospector-geologist, discoverer of Shinkolobwe uranium deposit, 1915.

SIMON, Francis: German-born physicist who fled Nazi racial laws for Oxford where he led the team which made first studies of separation of uranium 235 in gaseous diffusion.

SODDY, Frederick: With Rutherford, in Montreal, 1902, discovered spontaneous disintegration; proposed existence of isotopes; took part in discovery of natural decay chain of uranium and proved lead to be the final link.

STRASSMANN, Fritz: Collaborated with Hahn, late 1938, in Berlin on uranium experiment which led to the fission explanation.

SUFFOLK, Earl of: Scientific attaché to U.K. embassy in Paris in 1940; organized escape of scientists and world stock of heavy water, via Bordeaux.

SZILARD, Leo: Born in Budapest, became one of central characters of the uranium epic. In 1934, London, had notion of chain reaction; placed world's first nuclear patent with British Admiralty. Driving force behind uranium work at Columbia University and the Einstein letter to Roosevelt. Urged secrecy on research, 1939; in 1945 campaigned against use of bomb on Japan, and development of hydrogen weapons. Won Atoms for Peace Award in 1950.

TELLER, Edward: Hungarian physicist, worked in Hamburg, Berlin, Copenhagen, before fleeing to United States. With Szilard campaigned for U.S. uranium project. With Fermi first sketched the physical pathway to the hydrogen bomb, 1942. Later, made discovery of secret process to achieve fusion of light elements with fission explosion as the trigger. Widely known as "Father of the H-bomb."

THOMSON, J. J. (Sir): Cavendish professor, 1895, discoverer of the electron; set Rutherford on path of atomic discovery. Father of Sir George Thomson, chairman of M.A.U.D. Committee.

TITTERTON, Ernest: U.K. physicist, with Oliphant at Birmingham, cooperated with Frisch on fission measurements; first to notice spontaneous fission of uranium-238 atoms. British team member at Los Alamos, in charge of electronics for firing and monitoring plutonium explosion, New Mexico, July, 1945.

TIZARD, Henry (Sir): Scientific adviser to U.K. Air Ministry, guiding hand on radar development. Cast serious doubt on possibility of uranium bomb. Organized unrestricted flow of British scientific secrets to United States, 1941.

UREY, Harold: American chemist, in 1932 discovered Rutherford's projected two-part hydrogen, deuterium, which led to heavy water. Played advisory role in research by the Manhattan Project organization.

WAHL, Arthur C.: Student with Seaborg at Berkeley, 1940, worked on isolation and detection of element 94, plutonium. Later, worked on Manhattan Project.

WALTON, E. T. S.: In 1932, with Cockcroft, at Cavendish, built the first particle accelerator; in "splitting the atom" gave first demonstration of Einstein's law that mass and energy are equivalent.

WATTENBERG, Albert: Member of Fermi's team in Chicago, December, 1942, which built and brought to operation the first self-sustaining uranium reactor.

WEIZSÄCKER, Carl von: Son of Hitler minister; reputed to have worked on development of uranium reactor in Berlin, during the war.

WHEELER, John A.: Coauthor with Bohr of classic paper, 1939, describing the role of uranium-235 in fission process. Made pioneering contributions to understanding of nuclear reactions; took part in planning construction of huge reactors to produce plutonium for Manhattan Project.

WIGNER, Eugene: Hungarian who fled the Third Reich to America to play a leading role in early uranium research. One of those who urged Einstein to sign the letter to Roosevelt. Directed theoretical study in Manhattan Laboratories in Chicago and worked on reactor designs.

ZINN, Walter H.: American researcher in neutrons; at Columbia, 1939, with Szilard, observed chain reaction process. Worked with Fermi on first uranium reactor and became a leading American expert, responsible for first uranium reactor to produce electric power—known as "Zinn's Pile"—also designed first breeder reactor and first submarine propulsion unit. Became director, Argonne National Laboratory.

SOURCES

CODE

A.R.: Articles, reports; letters to journals.
O.H.R.: Official or historical records.
P.C.: Personal communication and correspondence.
P.P.L.: Private papers or letters.
P.S.: Published statements.
P.W.: Published works: books and biography.
R.I.: Recorded interviews.

FIRST PHASE

Berlin, 1780—Georg Dann *P.W.; Apotheker Zeitung A.R.*
Berlin, 1789—*Ibid.;* Prussian Academy annals; *P.P.L.*
Russia, 1872—British Library *P.W.;* Memorial Lectures *A.R.*
Germany, 1895—London Chemical Society *Lectures.*
Paris, 1896—Sir Oliver Lodge et al., British Library *A.R.*
Pungarehu—Eve Curie *P.W.;* Rutherford *P.P.L.;* Oliphant *P.C.*
Paris, 1903—Rutherford *P.P.L.;* Eve Curie *P.W.;* Mackown, *P.W.*
    Crowther *A.R.;* Nobel Institute *O.H.R.;* Eve Curie *P.W.;* Perrin
    A.R., O.H.R.
London, 1905—Robert Sharp *P.W.*
Frankfurt, 1905—Otto Hahn *P.W., A.R.*
Berne, 1905—Vallentin *P.W.;* Swinger, Clarke, Laurence *P.W., A.R.*
Montreal, 1906—Hahn *P.W.;* Rutherford, Soddy *A.R.*
Berlin, 1907—Otto Hahn *P.W.*
Manchester, 1911—Eve Curie *P.W., A.R., O.H.R.;* C. G. Darwin *A.R.*
Katanga, 1915—R. R. Sharp *P.W.*
Manchester, 1919—Chadwick, Eve Curie, Oliphant, *P. W.;* Crowther,
    A.R., O.H.R.
Katanga, 1919—Union Minière *O.H.R.;* R. R. Sharp *P.W.*
London, 1920—Royal Society *Proceedings;* Eve Curie, *P.W.;* Oliphant,
    Chadwick, *P.W., A.R.*
Belgium, 1920—Union Minière annals; Goldschmidt *P.W., A.R.;*
    Oliphant, Eve Curie, Chadwick *P.W.;* Cockcroft *P.C.*

SECOND PHASE

1—Oliphant *R.I.*, *P.C.*; Chadwick *P.C.*, *A.R.*; Eve Curie *P.W.*
2—Solvay Annals; Frisch *P.W.*; Biquard *P.W.*
3—Kowarski *R.I.*; *Journal de Physics A.R.*
4—Szilard *P.P.*, *P.W.*; Kowarski *R.I.*; Bainbridge *A.R.*, *P.S.*
5—Frisch *A.R.*, *R.I.*, *P.A.*, *O.H.R.*
6—Goldschmidt *P.W.*, *A.R.*; Segré *P.S.*; Seaborg *P.W.*, *A.R.*, *R.I.*; Laurence *P.W.*, *A.R.*; *Ricerca Scientifica A.R.*
7—Aage Bohr *P.W.*; Frisch *A.R.*, *R.I.*: Rozenthal *P.W.*
8—Dempster *P.W.*; Hewlett and Anderson *P.W.*; Hahn *P.W.*; Gregory *P.W.*
9—Frisch *R.I.*, *P.C.*; Rozental *P.W.*; Gamow *P.C.*
10—Kowarski, *P.C.*: Biquard *P.W.*; Joliot-Curie *A.R.*; Goldschmidt *P.C.*, *P.W.*
11—Frisch *P.C.*, *P.W.*; Hahn *P.W.*, *O.H.R.*
12—Frisch *R.I.*, *A.R.*; Rozental *P.W.*; A. Bohr *P.W.*; Laurence *A.R.*, *P.W.*; "All In Our Time," *A.R.*
13—Seaborg *R.I.*, *A.R.*, *P.W.*; *Science Service*; Laurence *P.W.*, *A.R.*, *O.H.R.*

THIRD PHASE

1—Kowarski *P.C.*, *R.I.*; Biquard *P.W.*, *O.H.R.*
2—Union Minière: Goldschmidt *P.W.*, *R.I.*; Sharp *P.W.*
3—Gowing *P.W.*, *O.H.R.*; Groves *P.W.*; U.K. war cabinet records.
4—A. Bohr *P.W.*; Rozental *P.W.*; Hewlett and Anderson *P.W.*; Frisch *R.I.*, *A.R.*; Hahn *P.W.*; Biquard *P.W.*; Kowarski R.I.; Churchill *P.W.*
5—Frisch *R.I.*, *A.R.*; U.K. cabinet records; Gowing *P.W.*, *O.H.R.*; Hewlett and Anderson *P.W.*
6—Kowarski *R.I.*, *A.R.*; Allier *R.I.* *P.S.*; A. Bohr *P.S.*, *A.R.*; U.K. cabinet records; Paris patents office.
7—Oliphant *R.I.*; Bohr *P.S.*; Gowing *P.W.*, *A.R.*; Frisch *R.I.*, *A.R.*; U.K. cabinet records.
8—Biquard *P.W.*; Kowarski *R.I.*, *A.R.*; Allier *R.I.*; Joliot-Curie *P.S.*; *O.H.R.*
9—Strauss *P.W.*; Union Minière *R.I.*, *P.C.*; Groves *P.W.*; Macmillan *P.W.*

10—O.H.R.: Biquard *P.W.;* Joliot-Curie *P.S.;* Kowarski *R.I.;* Frisch *R.I., P.S., A.R.;* Allier *R.I.;* Halban *P.S.;* Clarke *P.W.;* Union Minière; Hewlett and Anderson *P.W.*

11—U.K. war cabinet records; Gowing *P.W.;* U.K. Treasury statement: M.A.U.D. Records; Frisch *R.I., A.R.*

12—M.A.U.D. Documents; Gowing *P.W.;* Oliphant *R.I., P.S.;* Bainbridge *A.R.,* P.S.; "All In Our Time," *A.R.*

13—Gowing *P.W.;* U.K. war cabinet records; Oliphant *R.I.;* Cockcroft *P.S.*

14—Oliphant, *R.I.;* Pharr Davis *P.W.;* Laurence *P.W.;* A. Compton *P.W.;* Hewlett and Anderson *P.W.*

FOURTH PHASE

1—Hewlett and Anderson *P.W.;* Laurence *P.W., New York Times, A.R.;* H. Anderson *A.R.;* Segré *P.W.;* A. Bohr *A.R.;* Rozental *P.W.*

2—Strauss *P.W.;* Bush *P.W.;* Szilard *P.P.L;* Hewlett and Anderson *P.W.*

3—Szilard *P.S., P.P.L.;* Strauss *P.W.;* Hewlett and Anderson *P.W.;* Laurence *P.W.;* Teller *R.I., O.H.R.*

4—Hewlett and Anderson *P.W.;* Bush *P.W.;* Strauss *P.W.;* Szilard *P.P.L.;* Union Minière.

5—H. Anderson *A.R., P.S.;* Hewlett and Anderson, *P.W.; Physical Review A.R.; Science News A.R.;* Strauss *P.W.; O.H.R.;* Seaborg *R.I., P.C., P.W.*

6—Seaborg *R.I., A.R., P.W.;* Hewlett and Anderson *P.W.;* Seaborg *A.R.;* A. Compton *P.W.;* Gowing *P.W.;* H. Anderson *A.R.*

7—Seaborg *R.I., P.W.;* Kennedy *P.W.;* Berkeley records; Bainbridge *A.R.;* M.A.U.D. records; Oliphant *R.I.;* Hewlett and Anderson *P.W.*

8—A. Compton *P.W.;* Bainbridge *A.R.;* Pharr Davis *P.W.; O.H.R.;* Oliphant *R.I.; Nuclear Milestones A.R.;* U.K. war cabinet documents.

9—Strauss *P.W.;* Bush *P.W.;* Union Minière *R.I.;* Hewlett and Anderson *P.W.;* Gowing *P.W.;* Oliphant *R.I.;* M.A.U.D. records; Laurence *P.W.;* Griffith *P.W.;* Churchill *P.W.;* Groves *P.W.*

10—Groves *P.W.;* Bush *P.W.;* Hewlett and Anderson *P.W.;* A. Compton *P.W.;* Seaborg *R.I., A.R.; Nuclear Milestones,* U.S. AEC; Perlman *P.S.;* Werner *P.S., O.H.R.*

11—Groves *P.W.;* Bush *P.W.;* Union Minière *R.I.;*

12—H. Anderson *A.R., P.S.;* Segré *P.S.;* A. Compton *P.W.;* Hewlett and Anderson, *P.W.;* Conant *P.S.;* Wattenberg *A.R., O.H.R.*

13—Seaborg *O.H.R.*

FIFTH PHASE

1—*O.H.R.:* Groves *P.W.;* Hewlett and Anderson *P.W.;* Oppenheimer *P.S.;* Frisch *R.I., A.R.;* McDaniell *A.R.;* Jungk *P.W.;* Seaborg *P.S., P.W.*

2—A. Bohr *P.W.;* Rozental *P.W.;* Chadwick *P.S.;* Gowing *P.W.;* U.K. war cabinet records; Frisch *A.R.;* Groves *P.W.;* Strauss *P.W.;* Union Minière; Bush *P.W.*

3—*O.H.R.:* Groves *P.W.;* Los Alamos records; Serber *P.S.*

4—Rozental *P.W.;* A. Bohr P.W.; Frisch *R.I., A.R.;* Los Alamos history *P.W.;* Hewlett and Anderson *P.W.;* Groves *P.W.;* O.H.R.; Bainbridge A.R.; Irving *P.W.;* Hahn *P.W.*

5—Frisch *R.I., A.R.;* Los Alamos history *P.W.;* Bainbridge *P.S., A.R.*

6—Bainbridge *A.R.;* H. Anderson *P.S., A.R.;* Hewlett and Anderson *P.W.;* Frisch *A.R.;* Los Alamos history *P.W.;* McDaniell, *A.R.;* Fitch *A.R.*

7—Frisch *R.I., A.R.;* Bainbridge *A.R., P.S.*

8—Los Alamos history *P.W.;* Seaborg *P.C.;* Farrell *A.R.;* Frisch *R.I.;* Hewlett and Anderson *P.W.;* Gowing *P.W.;* H. Anderson *A.R.;* Bainbridge *A.R.*

9—Los Alamos history *P.W.;* Groves *P.W.;* Hewlett and Anderson *P.W.;* Manhattan history "Project Y" *P.W.;* Frisch *R.I., A.R.;* Strauss *P.W.;* Groves *P.W.;* Sengier *A.R.;* *"Human Events"* *A.R.;* Sharp *P.W.*

TRANSMUTATION

O.H.R.; Churchill; Hewlett and Duncan *P.W.*

U.N. records; Blackett *P.W.;* Laurence *P.W.*

Cameron *A.R.;* E. Lawrence *P.S.*

Nuclear Milestones, U.S. AEC *A.R.;* Goldschmidt *P.W.*

Gowing *P.W.;* U.N. Resources Conference; Kamish *P.W.*

Price *Uranium Institute Report;* Biew *P.W.;* R. Lapp *P.W.*

Seaborg *A.R.;* Union Minière records: Oliphant *R.I.*

Seaborg *R.I.;* Frisch *R.I.*

BIBLIOGRAPHY

BOOKS

Atomic Adventure: Bertram Goldschmidt; Commonwealth & International Library.

Atomic Energy in the Soviet Union: Arnda Kramish; Stanford University Press.

Atomic Physics Today: Oliver & Boyd; London, 1962.

Atomic Quest: Arthur Compton; Oxford University Press.

Atomic Shield: Hewlett and Duncan; U.S. Atomic Energy Commission.

Birth of the Bomb: Ronald Clarke; London.

Brighter Than a Thousand Suns: Robert Jungk; Gollancz, London, 1958.

Britain and Atomic Energy, 1939-1945: Margaret Gowing; Macmillan, U.K.

The Curve of Binding Energy: John McPhee; Ballantine, N.Y.

Discovery of the Elements: H. M. Leicester; London, 1968.

Early Days in Katanga: R. R. Sharp; Rhodesian Printers, Bulawayo, Southern Rhodesia.

Early History of Heavy Isotopes: Glenn T. Seaborg; University of California.

Einstein: Antonina Vallentin; Weidenfeld & Nicolson, London.

German Chemical Community: Karl Haufbauer; University of California, 1970.

Halley Stewart Lectures: Oliphant-Blackett-Russell; London Chemical Society.

History of Met. Lab., Section C-1: Glenn T. Seaborg; University of California.

Joliot-Curie: Pierre Biquard; Souvenir Press, London, 1965.

Kapitsa: A. M. Biew; Frederick Muller & Co., London, 1956.

Kill and Overkill: Ralph Lapp; Basic Books, N.Y.

Lawrence: An American Genius: Nuel Pharr Davis. New York.

The Laws of Nature: R. Peierls; Thames & Hudson, London.

Le Radium et les Radio-Éléments (Preface, Marie Curie): Baillière et fils, Paris, 1925.

Los Alamos Primer: Serber; U.S. Department of Commerce.

L'uranium du Congo: C. d'Ydewalle; Éditions L. Cuypers, Brussels.

Manhattan District History: (Los Alamos Project) U.S. Department of Commerce.

The Manhattan Project: S. Grueff; Little, Brown, Boston.

Martin Heinrich Klaproth: Georg Dann; Akademie-Verlag, Berlin, 1958.

Men and Atoms: W.L. Laurence; Simon & Schuster, New York.

Men and Decisions: Lewis Strauss; Doubleday, N.Y., 1962.

Modern Inorganic Chemistry: J.W. Mellor; Longmans.

The Nature of Matter: Otto R. Frisch; Thames & Hudson, London.

The New World: Hewlett and Anderson, U.S. Atomic Energy Commission.

Niels Bohr—His Life and Work: S. Rozental, ed.; Wiley, New York.

Niels Bohr—The Man and His Science: Ruth Moore; Knopf, N.Y.

Now It Can Be Told: Leslie Groves; Harper, New York.

Nuclear Fuels Policy: Atlantic Institute, New York.

Nuclear Milestones: G. T. Seaborg; U.S. Atomic Energy Commission.

Nuclear and Radio Chemistry: Friedlander-Kennedy-Miller; Wiley, New York.

Passion To Know: Mitchell Wilson; Doubleday, New York.

Pieces of the Action: Vannevar Bush; Morrow, New York, 1970.

Plutonium Handbook (Vols. 1 & 2): U.S. Atomic Energy Commission.

Recollections of Cambridge Days: M. L. Oliphant; Elsevier, London, 1972.

Reserves of Uranium in U.S.A.: U.S. Academy of Sciences.

Rutherford: A. S. Eve; London.

Rutherford: R. Mackown; A. & C. Black, London.

Rutherford and the Nature of the Atom: E. N. da C. Andrade; London.

Rutherford—Master of the Atom: J. Rowland, London.

Rutherford Papers: J. Chadwick, ed.; London.

A Scientific Biography: Otto Hahn; Scribner, New York.

The Secret History of the Atomic Bomb: Brown-MacDonald; Delta, N.Y.

The Second World War: Winston Churchill; Bantam Books, N.Y.

She Lived for Science: R. Mackown; Macmillan, U.K.

Transuranic Elements: G. T. Seaborg; Yale University (Silliman Lectures, Vol. 37).

Uranium: J. H. Gittus; Butterworths, London, 1963.

The Uranium Industry—Canada: J. W. Griffith; Dept. of Energy, Ottawa.

Uranium and Thorium: L. Grainger; George Newnes, London.

The Virus House: David Irving; Kimber, U.K.

The World of Radio Isotopes: J.N. Gregory; Angus & Robertson, Sydney.

PAMPHLETS AND MAGAZINES

All in Our Time: Jane Wilson et al., eds.; *Bulletin of the Atomic Scientists.*

"How It All Began": O. Frisch; *Physics Today* November, 1967.

MAUD Reports: U.K. Govt. Cabinet Papers: File 1262X/K5554.

"Disintegration of Uranium": Otto Frisch and Lise Meitner, *Nature,* 1939.

Enrico Fermi: Last Address; *Physics Today;* U.S. Physical Society, 1954.

Nuclear Energy in Britain: U.K. Central Office of Information.

"Nuclear Power—How Dangerous?": Ralph Lapp; *Reader's Digest.*

The Atomic Bomb: U.K. Treasury, 1939.

"Otto Frisch": *New Scientist,* May, 1974; *Nature,* March, 1939.

"Joliot-Curie et al.": *Nature,* April, 1939.

Interest on the Atomic Nucleus: North-Holland, 1967.

Products of Fission of Uranium: Meitner-Frisch; Ejnar Munksgaard, Copenhagen, 1939.

"Nuclear Fusion": M. H. Brennan, *Search,* Sydney.

London Chemical Society Lectures:—Becquerel Memorial (Sir Oliver Lodge, 1914): "Mendeleev, Rutherford And The Nature of the Atom," Andrade; Cawthorne Lectures, "Memories of Rutherford"; Sir Henry Dale.

"Rutherford the Great": J. G. Crowther, *New Scientist,* August, 1971.

"Eine Deutsche Apotheke": *Österreichische Apotheker Zeitung,* January, 1955.

Uranium and Thorium Geology: U.N. International Conference on Atomic Energy, Geneva, 1955.

ACKNOWLEDGMENTS

This narrative was written with the generous cooperation of many people, and from the records of various organizations. So wide has been the support, over several years of research, that the point must be made: Order of mention does not necessarily indicate the value of assistance.

Nonetheless, Sir Mark Oliphant, of Canberra, must head the list; he has been a font of data and encouragement, and he has now capped this support in the many hours spent reading the manuscript. I am deeply grateful to Sir Mark for his suggestions to improve accuracy of expression. I also express my gratitude to Dr. Glenn T. Seaborg, of Berkeley, California, for many hours spent in interviews, and for access to copies of his diaries written during the hectic years of war; and to Professor Otto Robert Frisch, of Cambridge, England, for exceeding generosity with time, recollections, records, and documents. I am also indebted to Professor Ross Taylor, of Canberra, for scanning the chemistry details; to Professor Lew Kowarski, of Geneva, for recorded interviews, and his observations of the Joliot-Curie work; to the late Sir John Cockcroft and Sir James Chadwick; and to Professor Charles Watson-Munro, of Sydney. Conversations with Dr. Edward Teller, Sir Ernest Titterton, and the late Dr. George Gamow were very helpful.

From the Union Minière Du Haut Katanga organization, I have had strong support; from the Administrator-Director, M. Joseph Derriks, the Vice-President, M. A. Assoignon, and senior geologists A. Kazmitcheff and Kenneth Glasson. The aid of Dr. Bertram Goldschmidt, of Paris, was invaluable, as was that of officials of the Collège de France, and the former secret service officer, M. Jacques Allier. And I sincerely acknowledge the contributions made to this work by Dr. John Brooke, of Lancaster

University; Professor Margaret Gowing, the nuclear historian; Dr. Terence Price, of the London Uranium Institute; Dr. Richard Hewlett, of Washington; and Mr. J. St. Cloud, of the U.K. AEA.

Additional travel and research was made possible by the support of the Literature Board of the Australia Council; and officers of various organizations were unstinting in their help: Egbert Keiser and his research staff of *Das Beste,* of Stuttgart, and Dimi Panitzer and his aides of the same organization, in Paris. The support and aid of my wife, Pauline, was indispensable; and I am deeply indebted to librarians, especially Anne Fraser of Sydney's *Reader's Digest,* and those of the Australian National Library, the British Library, and various universities in different lands. Also, I was greatly assisted by having access to documents from the Public Records Office, London, from U.S. government archives, and from the Royal Society.

Conversations in this work are drawn directly from written accounts, private papers, diaries, and from the memories of those living who take part in the story.

In the cause of easy reading, scientific terminology—which Winston Churchill found ". . . so incomprehensible to ordinary ears"—has been reduced to simpler expression. If scientific pedants find more abstruse points of atomic physics have been avoided, then I accept responsibility since it has been done to add freshness to the drama of these historic events.

LENNARD BICKEL

INDEX